Governing through Goals

Earth System Governance

Frank Biermann and Oran R. Young, series editors

Oran R. Young, *Institutional Dynamics: Emergent Patterns in International Environmental Governance*

Frank Biermann and Philipp Pattberg, eds., *Global Environmental Governance Reconsidered*

Olav Schram Stokke, *Disaggregating International Regimes: A New Approach to Evaluation and Comparison*

Aarti Gupta and Michael Mason, eds., *Transparency in Global Environmental Governance: Critical Perspectives*

Sikina Jinnah, *Post-Treaty Politics: Secretariat Influence in Global Environmental Governance*

Frank Biermann, *Earth System Governance: World Politics in the Anthropocene*

Walter F. Baber and Robert B. Bartlett, *Consensus in Global Environmental Governance: Deliberative Democracy in Nature's Regime*

Diarmuid Torney, *European Climate Leadership in Question: Policies toward China and India*

David Ciplet, J. Timmons Roberts, and Mizan R. Khan, *Power in a Warming World: The New Global Politics of Climate Change and the Remaking of Environmental Inequality*

Simon Nicholson and Sikina Jinnah, eds., *New Earth Politics: Essays from the Anthropocene*

Norichika Kanie and Frank Biermann, eds., *Governing through Goals: Sustainable Development Goals as Governance Innovation*

Related books from Institutional Dimensions of Global Environmental Change: A Core Research Project of the International Human Dimensions Programme on Global Environmental Change

Oran R. Young, Leslie A. King, and Heike Schroeder, eds., *Institutions and Environmental Change: Principal Findings, Applications, and Research Frontiers*

Frank Biermann and Bernd Siebenhüner, eds., *Managers of Global Change: The Influence of International Environmental Bureaucracies*

Sebastian Oberthür and Olav Schram Stokke, eds., *Managing Institutional Complexity: Regime Interplay and Global Environmental Change*

Governing through Goals

Sustainable Development Goals as Governance Innovation

edited by Norichika Kanie and Frank Biermann

The MIT Press
Cambridge, Massachusetts
London, England

This book was set in Stone Sans and Stone Serif by Toppan Best-set Premedia Limited. Printed on recycled paper and bound in the United States of America.

Library of Congress Cataloging-in-Publication Data

Names: Kanie, Norichika, 1969- editor. | Biermann, Frank, 1967- editor.
Title: Governing through goals : sustainable development goals as governance innovation / edited by Norichika Kanie and Frank Biermann.
Description: Cambridge, MA : MIT Press, [2017] | Series: Earth system governance | Includes bibliographical references and index.
Identifiers: LCCN 2016026443| ISBN 9780262035620 (hardcover : alk. paper) | ISBN 9780262533195 (pbk. : alk. paper)
Subjects: LCSH: Sustainable development--International cooperation. | Environmental policy--International cooperation. | Millennium Development Goals. | Sustainable Development Goals.
Classification: LCC HC79.E5 .G6675 2017 | DDC 338.9/27--dc23 LC record available at https://lccn.loc.gov/2016026443

10 9 8 7 6 5 4 3 2 1

Contents

Series Foreword

Humans now influence all biological and physical systems of the planet. Almost no species, land area, or part of the oceans has remained unaffected by the expansion of the human species. Recent scientific findings suggest that the entire earth system now operates outside the normal state exhibited over at least the past 500,000 years. At the same time, it is apparent that the institutions, organizations, and mechanisms by which humans govern their relationship with the natural environment and global biogeochemical systems are utterly insufficient—and poorly understood. More fundamental and applied research is needed.

Such research is no easy undertaking. It must span the entire globe, because only integrated global solutions can ensure a sustainable coevolution of biophysical and socioeconomic systems. But it must also draw on local experiences and insights. Research on earth system governance must be about places in all their diversity, yet seek to integrate place-based research within a global understanding of the myriad human interactions with the earth system. Eventually, the task is to develop integrated systems of governance, from the local to the global level, that ensure the sustainable development of the coupled socioecological system the Earth has become.

This series, Earth System Governance, is designed to address this research challenge. Books in this series will pursue this challenge from a variety of disciplinary perspectives, at different levels of governance, and with a range of methods. Yet all will further one common aim: analyzing current systems of earth system governance with a view to increased understanding and possible improvements and reform. Books in this series will be of interest to the academic community, but will also inform practitioners and at times contribute to policy debates.

This series is related to the long-term international research program, the Earth System Governance Project.

Frank Biermann, Copernicus Institute of Sustainable Development, Utrecht University
Oran R. Young, Bren School, University of California, Santa Barbara
Earth System Governance Series Editors

Preface

It was in September 2011, in the midst of the Hakone Mountains in Japan, when we first learned about the novel idea of "sustainable development goals." This idea was advanced by Jimena Leiva Roesch, then representative of the government of Guatemala to the United Nations, at an intense brainstorming workshop called the Hakone Vision Factory: Bridging Science Policy Boundaries, hosted by the Earth System Governance Project and the International Environmental Governance Architecture Research Group. This "vision factory" originally focused on the institutional framework for sustainable development as one of the two main themes of the 2012 UN Conference on Sustainable Development; yet our discussions quickly expanded to include new, broader visions for sustainable development governance in the twenty-first century. The idea of "sustainable development goals" was both striking and enlightening, and a consensus quickly emerged among participants that promoting this idea would benefit governance for sustainability. Yet nobody at that point imagined that the sustainable development goals would attract greatest attention in the UN debate soon thereafter.

The evolution of the sustainable development goals over the last four years is truly phenomenal, and we feel privileged to have been able to witness this development. The more we learned about the sustainable development goals, the more we realized their potential and innovative implications. Governing through goals is fundamentally different in character from other forms of global governance. It is a new governance strategy that has emerged at an opportune time under the novel conditions of the twenty-first century. And as such, it requires careful academic investigation.

This first book on global governance through goal setting is a product of many different but interconnected projects about governance and sustainability. The first of our workshops to discuss the governance implications of sustainable development goals took place during the 2013 Earth System

Governance Tokyo Conference. The conference was supported by the Japan Foundation Center for Global Partnership, the Institute for Global Environmental Strategies in Japan, the Paris-based Institut du Développement Durable et des Relations Internationales, the Japan Science and Technology Agency, and Japan's Ministry of the Environment through its Global Environmental Research Fund (RFe-1201). After intense discussions on possible new and innovative dimensions of governance of, as well as for, the sustainable development goals, participants became interested in the idea to launch a new project specifically to look at the governance dimensions of the sustainable development goals. Participants in the first workshop included, among others, Steinar Andresen, Joyeeta Gupta, Peter M. Haas, Marc Levy, Måns Nilsson, László Pintér, Laurence Tubiana, and Takahiro Yamada.

In April 2013, one of us—Norichika Kanie—was commissioned to lead a strategic research project ("S-11") of the Environment Research and Technology Development Fund of Japan's Ministry of the Environment. This project, entitled "Project on Sustainability Transformation beyond 2015," or POST2015, was designed to investigate both human and planetary wellbeing and how both could be consolidated in the making and implementation of sustainable development policies, with the primary aim of providing inputs and evidence for international discussions on a post-2015 development agenda and the sustainable development goals. This book project has been part of the wider framework of this strategic research project, through the leadership team and one of the subthemes commissioned at the United Nations University Institute for the Advanced Study of Sustainability (until 2013, the United Nations University Institute of Advanced Studies). Thus, our sincere thanks are due to the Ministry of Environment of Japan; and as project leader, Norichika Kanie wishes to mention in particular the following persons who have helped realize the policy-oriented project: Ryutaro Yatsu, Soichiro Seki, Yutaka Matsuzawa, Hiroshi Tsujihara, Akio Takemoto, Naoya Tsukamoto, Keiko Segawa, Atsushi Takenaka, Yasukuni Shibata, Shuichi Mizushima, Yuta Higuchi, Takuya Fusamura, Maya Suzuki, and Hirofumi Karibe.

This governance project was conducted through the UN University Institute for the Advanced Study of Sustainability, in close collaboration with the Earth System Governance Project, the leading international transdisciplinary network in the field of sustainability governance studies. All authors in this book have been involved in the Earth System Governance Project in one way or another, as members of its Lead Faculty, research fellows, participants in global conferences, or as members of the project's

Scientific Steering Committee, chaired by Frank Biermann. Two colleagues deserve special acknowledgment. First, Oran R. Young has been closely involved in, and very supportive of, this collaborative effort from its start. Without his support, this project would not have attracted the participation of the many outstanding scholars who contributed chapters in this book. Another person who has provided unconditional and unlimited support is Ruben Zondervan, the executive director of the Earth System Governance Project, without whom this project would not have materialized. Ruben has participated in all the workshops and meetings and provided numerous and highly useful comments. We also thank Lund University, Sweden, for hosting the Earth System Governance International Project Office, and for generously providing the infrastructure and support for a number of interns who supported this endeavor. We thank here especially the many highly talented and enthusiastic interns who have worked on the sustainable development goals at the Earth System Governance International Project Office in Lund, notably Javier Munoz-Blanco, Maria Dahlman Ström, Henry Kröger, Johannes Nilsson, and Jonathan Volt.

The kick-off workshop of the book project was held in July 2013 as part of the S-11 project inaugural meeting in Yokohama, Japan, chaired by Norichika Kanie. A first outline of the book project was elaborated, and potential themes and contributors were identified. At the workshop the ideas behind launching the book project were elaborated, and initial ideas of the issues to be addressed in each chapter were discussed. An important element of the process was that wherever possible, senior and junior researchers were brought together in writing chapters, with the aim of making this research community sustainable and combining the cutting-edge knowledge and inspiration of both the younger and the more advanced generations. About half a year later we organized a subsequent International Workshop on Governance of, and for, Sustainable Development Goals in New York. At that time, the sustainable development goals received much more political attention because the formal negotiation phase had begun. Our workshop was thus explicitly organized as a transdisciplinary effort that brought together both researchers and practitioners in a "world café" discussion format to generate novel ideas on the governance dimensions of the sustainable development goals.

The workshop was held just before the eighth meeting of the UN General Assembly's Open Working Group on sustainable development goals, which concluded the "stock-taking" phase of this group. One major topic then was the architecture of the sustainable development goals—that is, how to formulate the sustainable development goals at the global level while securing diversity of implementation in different national circumstances. Although our workshop did not directly discuss the contents of this book, some outcomes of the workshop—especially the policy briefs resulting from the meeting—formed the basis of some chapters of this book. Hence, all participants in the workshop who are not formally contributors of chapters to this volume deserve full acknowledgment for their input, starting with Csaba Kőrösi, a co-chair of the Open Working Group, and including Mathilde Bouyé, Guy Brendan, Olivia Caeymaex, Ngeta Kabiri, Yuto Kitamura, Maja Messmer Mokhtar, Ian Noble, Simon Høiberg Olsen, Barbara J. Ryan, Mayumi Sakoh, Masahisa Sato, Anne-Sophie Stevance, Jan-Gustav Strandenaes, Zoltán Szentgyörgy, Farooq Ullah, Peter Veit, Dongyong Zhang, Janos Zlinszky, and Irena Zubcevic.

The first draft of this book was then discussed at two authors' workshops, in Toronto in March 2014 and in Paris in September 2014. It was at the time of the Paris workshop that the Open Working Group outcome of 17 goals and 169 targets was decided at the sixty-ninth session of the UN General Assembly as the main input to the post-2015 development agenda, and the now-crucial importance of this new governance strategy became even more evident.

Earlier versions of this book and several chapters were presented at various occasions, including the conference "Our Common Future under Climate Change" in Paris in July 2015; the annual convention of the International Studies Association in New Orleans in February 2015; the convention of the Society for Environmental Economics and Policy Studies in Kobe in September 2013; and the fourteenth conference of the Japan Society for International Development in June 2013. We would like to extend our appreciation for all comments and suggestions provided at these occasions, which have improved many core ideas included in our conceptual development.

Apart from the aforementioned persons, we have received useful advice from practitioners, academics, and those who take on the difficult task of going in between both. Norichika Kanie, in particular, as overall project leader, wishes to thank very much Keizo Takemi, a member of the House of Councilors, along with Sakiko Fukuda-Parr, David Griggs, Shinishi Iida, Keita Iwase, Takehiro Kagawa, Kaori Kuroda, Hiroshi Minami, Shuzo Nishioka, Atsuyuki Oike, Tomoko Onishi, Atsushi Suginaka, Motoyuki Suzuki, Naohito Asano, Katsunobu Takada, Kazuhiro Takemoto, Kazuhiko Takeuchi, Masami Tamura, Ikufumi Tomimoto, and Kazuhiro Ueta.

Project partners from the International Institute for Sustainable Development have given us additional valuable information on the internal details of the UN workings and beyond, and we thank in particular Langston James "Kimo" Goree VI, Pamela Chasek, Lynn Wagner, Faye Leone, and Kate Offerdahl. Special thanks is due to Csaba Kőrösi, Ambassador and Permanent Representative of Hungary to the United Nations and a co-chair of the Open Working Group; Janos Zlinszky, Special Advisor to the Open Working Group Co-chair; and David O'Connor and Richard A. Roehrl from the UN Division for Sustainable Development, Department of Economic and Social Affairs.

Furthermore, all chapter authors would like to thank their individual supporters and funding agencies, which have made this project possible. In particular, László Pintér, Marcel Kok, and Dora Almassy would like to acknowledge three additional reviews for their chapter by Peter Bartelmus, professor at the Bergische Universität Wuppertal; Marianne Beisheim, senior associate with the German Institute for International and Security Affairs; and Steven Bernstein, who is also a contributor to this book. Steinar Andresen and Masahiko Iguchi acknowledge support from the International Collaboration for Capitalizing on Cost-effective and Life-saving Commodities under the Global Health and Vaccine Program of the Norwegian Research Council. Takahiro Yamada would like to thank Gavin Power, deputy executive director at the UN Global Compact, for generously spending two hours in an interview. He would also like to extend his appreciation to members of the Orchestration Project Team, funded by the Japan Society for the Promotion of Science, for comments on the draft chapter when he gave the paper at its meeting on May 7, 2015: Prof. Makiko Nishitani of Kobe University; Prof. Satoshi Miura of Nagoya University; and Prof. Yoshiko Naiki of Osaka University. Steven Bernstein would like to thank David O'Connor and Irena Zubcevic at the UN Department of Economic and Social Affairs for invitations to contribute to the UN's work on institutional reform following the 2012 UN Conference on Sustainable Development, from which many of the ideas for chapter 9 first evolved, as well as their valuable feedback on that research. He would also like to thank Kenneth Abbott for their joint work on orchestration that informs parts of chapter 9. Portions of his work in this volume were presented to expert group meetings sponsored by the United Nations and groups of governments and civil society organizations, including "Friends of Governance for Sustainable Development," and to academic workshops including "Rio+20 to 2015: A New Architecture for a Sustainable New World" at Yale's School of Forestry and Environmental Studies (November 2013) and the "Arizona Workshop on Implementing the Sustainable Development Goals," at University of Arizona in April 2015, all of which provided valuable discussions, context, and feedback. Måns Nilsson would like to thank for financial support the Swedish International Development Cooperation Agency. Joyeeta Gupta would like to thank the support of the Amsterdam Institute of Social Science Research at the University of Amsterdam, and the UNESCO-IHE Institute for Water Education. She also benefited from comments received when she presented "Governing the Risks with Respect to the Millennium Development Goal on Water, Water and Sustainable Development" at the 2015 UN Water Annual International Zaragosa Conference in Zaragosa in January 2015. Casey Stevens would like to acknowledge financial support from the Japan Society for the Promotion of Science and his colleagues at the United Nations University Institute for the Advanced Study of Sustainability.

Furthermore, Norichika Kanie wishes to extend special thanks to all members of his research team, notably Yuri Akita, Kaori Eto, Rieko Horie, Mie Iijima, Akemi Porter, Hitomi Shimatani, Kyoko Suganoya, Chiharu Takei, Noriko Takemura, Chikako Tokuda, as well as his research assistants, Yuka Hayakawa, Ikuho Miyazawa, and Yui Nakagawa. Mari Kosaka and Maki Koga should be particularly mentioned for their devotion to finalization of the project. Without their enthusiasm and dedication, the project would have been much more difficult!

And of course, thanks are due to all the contributors to this book. It has been enjoyable, inspirational, and, indeed, lots of fun to have been able to work with them!

Many thanks also to the excellent team at MIT Press. First of all, we are grateful to the anonymous reviewers, who helped to improve the manuscript with numerous highly constructive comments. Second, we wish to thank Beth Clevenger for her support, encouragement, and useful advice during the smooth and efficient review and production process of this book. Thirdly, we wish to thank the copyeditor, Kristie Reilly.

Last but not least, we wish to thank our families. Norichika Kanie, as the overall leader, had to take on extraordinary travel duties in managing and presenting this broader project. He wishes to thank in particular his wife, Reiko, for her support and understanding of his work. At the first project workshop in Hakone, in 2011, Reiko brought with her the youngest participant to the workshop, Hugo, who was then a 10-month old baby, and hence possibly the youngest person ever to have heard about the sustainable development goals. Hugo is now five years old at the time of conclusion of this book, and he will be 20 years old in 2030, when the sustainable development goals will hopefully have been achieved. The future is in his, and his generation's, hands. This book is, therefore, for all our children, who will grow up in the age of "governing through goals."

Norichika Kanie and Frank Biermann
Fujisawa, Japan, and Utrecht, The Netherlands
March 2016

List of Acronyms

ECOSOC	United Nations Economic and Social Council
FAO	UN Food and Agriculture Organization
GAVI	Global Alliance for Vaccines and Immunization
GDP	Gross domestic product
GNP	Gross national product
OECD	Organisation for Economic Co-operation and Development
SDSN	Sustainable Development Solutions Network
UN	United Nations
UNCED	United Nations Conference on Environment and Development
UNDP	United Nations Development Programme
UNECE	United Nations Economic Commission for Europe
UNEP	United Nations Environment Programme
UNESCO	United Nations Educational, Scientific and Cultural Organization
UNGA	United Nations General Assembly
WHO	World Health Organization
WTO	World Trade Organization

1 Introduction: Global Governance through Goal Setting

Norichika Kanie, Steven Bernstein, Frank Biermann, and Peter M. Haas

In September 2015, the UN General Assembly adopted the Sustainable Development Goals as an integral part of the 2030 Agenda for Sustainable Development (UNGA 2015). The Sustainable Development Goals were to build upon and broaden the scope of the earlier Millennium Development Goals, which had expired in the same year. The Sustainable Development Goals mark a historic shift for the United Nations toward one "sustainable" development agenda after a long history of trying to integrate economic and social development with environmental sustainability. They also mark the most ambitious effort yet to place goal setting at the center of global governance and policy. Governments' enthusiasm for goal setting, however, is not yet matched by knowledge about its prospects or limits as a governance strategy. This book aims to address this knowledge gap through a detailed examination of the Sustainable Development Goals and the governance challenges they face.

Neither goal setting nor sustainability are new approaches to world politics, development, or earth system governance. The United Nations, among other grand historical projects, is firmly rooted in broader goals such as justice, equality, and peace (or the elimination of war). Goal-setting has also been a feature of many multilateral agreements and programs of international institutions (Ruggie 1996; Williams 1998). Meanwhile, "sustainable development" and "sustainability" served as the conceptual cornerstones for the 1992 UN Conference on Environment and Development ("Rio Earth Summit"), the 2002 World Summit on Sustainable Development, and the 2012 UN Conference on Sustainable Development ("Rio plus 20"). But the Sustainable Development Goals go a step further than these earlier efforts. They add detailed content to the concept of sustainable development, identify specific targets for each goal, and use the concept to help frame a broader, more coherent, and transformative 2030 agenda.

This single, goal-oriented agenda is not simply a continuation of unfinished elements of the Millennium Development Goals; it aspires to build from their central mission of poverty eradication and social inclusion a universal, integrated framework for action that also responds to growing economic, social, and planetary complexity in the twenty-first century. Some may wonder whether goal seeking is a deliberate effort to evade the sort of commitments that were developed after the fact for the Millennium Development Goals. Others have questioned whether the particular formulation of sustainable development in the Sustainable Development Goals provides a sufficient foundation for a comprehensive agenda that includes human rights, social and political inclusion, and good governance (Browne 2014). The combination of extraordinary ambition, uncertain political commitments, and questions about the ability of goals to mobilize political and economic actors, and the resources required to pursue them, motivates three sets of questions that animate this volume.

First, the book studies in detail the core characteristics of goal setting in global governance, asking when it is an appropriate strategy in global governance and what makes global governance through goals different from other approaches such as rule making or norm promotion. Second, the chapters analyze under what conditions a goal-oriented approach can ensure progress toward desired ends; what can be learned from other, earlier experiences of global goal setting, especially the Millennium Development Goals; and what governance arrangements are likely to facilitate progress in implementing the new Sustainable Development Goals. Third, the book studies the practical and operational challenges involved in global governance through goals in promoting sustainability and the prospects for achieving such a demanding new agenda.

While these questions inform all chapters in this volume to varying degrees, chapters 2 to 5 focus especially on the first question. Chapters 6–8 most directly address the second question. Chapters 9–12, on operational challenges of goal attainment and implementing the Sustainable Development Goals globally and nationally, primarily focus on the third question.

Apart from advancing sustainable development worldwide, the Sustainable Development Goals are also an important focus of study in their own right as a new type of global governance. The perceived success of the Millennium Development Goals (an evaluation that a number of chapters in this book critically assess) has set the stage for this elevation of goal setting as a governance strategy. The Sustainable Development Goals now raise the stakes for this strategy, in part owing to the very public and high-level political process that has produced them.

Although the Millennium Development Goals reflected outcomes from many earlier UN and other international processes, as well as consultations with governments and UN agencies before and after the 2000 Millennium Summit, their specific formulation came from the UN Secretariat (McArthur 2014). The eight concise, yet broad, Millennium Development Goals and attendant targets were not negotiated outcomes (see this volume, Annex 1). In contrast, the Sustainable Development Goals required over two years of intense intergovernmental stocktaking and negotiation sessions, and perhaps the largest public and multi-stakeholder consultations in UN history. They are not simply standalone goals, but form the centerpiece of the broader new UN agenda approved by the UN General Assembly in September 2015: "Transforming Our World: The 2030 Agenda for Sustainable Development" (UNGA 2015; see table 1.1 for a list of the goals).

This encompassing declaration also reflects its own extensive negotiating and consultation process and incorporates the outcomes of numerous related international processes, including the third International Conference on Financing for Development (UN 2015; Voituriez et al., this volume, chapter 11) and the third World Conference on Disaster Risk Reduction, both held earlier in 2015. It even includes a space for the then forthcoming outcome from the twenty-first session of the Conference of the Parties to the United Nations Framework Convention on Climate Change, which is now called the Paris Agreement. The UN Secretary-General's synthesis report on a variety of inputs provided for the post-2015 development agenda was published just before the start of the final intergovernmental deliberation in 2015, which aimed to create a vision around which these various streams could cohere (UN 2014b).

The 2030 Agenda for Sustainable Development also pays attention to the means of delivering on its ambition, recognizing that achieving the Sustainable Development Goals will require not only a broader effort through the UN system, but also the mobilization of political support and resources well beyond it, including at regional and national levels and among multiple civil society, financial, and business actors. In sum, as the agreed title of the wider 2030 Agenda for Sustainable Development reflects, the Sustainable Development Goals aim at "transforming our world."

In the remainder of this chapter, we lay out a research agenda to assess conditions, challenges, and prospects for the Sustainable Development Goals to pursue this aim. First, we discuss goal setting as a global governance strategy. Second, to contextualize the Sustainable Development Goals, we discuss the unique nature of the contemporary challenges that

Table 1.1
Final List of Sustainable Development Goals

Goal 1	End poverty in all its forms everywhere
Goal 2	End hunger, achieve food security and improved nutrition, and promote sustainable agriculture
Goal 3	Ensure healthy lives and promote well-being for all at all ages
Goal 4	Ensure inclusive and equitable quality education and promote lifelong learning opportunities for all
Goal 5	Achieve gender equality and empower all women and girls
Goal 6	Ensure availability and sustainable management of water and sanitation for all
Goal 7	Ensure access to affordable, reliable, sustainable, and modern energy for all
Goal 8	Promote sustained, inclusive, and sustainable economic growth, full and productive employment, and decent work for all
Goal 9	Build resilient infrastructure, promote inclusive and sustainable industrialization, and foster innovation
Goal 10	Reduce inequality within and among countries
Goal 11	Make cities and human settlements inclusive, safe, resilient, and sustainable
Goal 12	Ensure sustainable consumption and production patterns
Goal 13	Take urgent action to combat climate change and its impacts*
Goal 14	Conserve and sustainably use the oceans, seas and marine resources for sustainable development
Goal 15	Protect, restore, and promote sustainable use of terrestrial ecosystems, sustainably manage forests, combat desertification, and halt and reverse land degradation and halt biodiversity loss
Goal 16	Promote peaceful and inclusive societies for sustainable development, provide access to justice for all, and build effective, accountable, and inclusive institutions at all levels
Goal 17	Strengthen the means of implementation and revitalize the Global Partnership for Sustainable Development

* Acknowledging that the UN Framework Convention on Climate Change is the primary international, intergovernmental forum for negotiating the global response to climate change.

Source: UN General Assembly. 2015. Transforming Our World: The 2030 Agenda for Sustainable Development. Draft resolution referred to the UN summit for the adoption of the post-2015 development agenda by the General Assembly at its sixty-ninth session. UN Doc. A/70/L.1 of September 18.

the Sustainable Development Goals must confront and review the historical and political trajectory of sustainable development governance, including the evolution from a primarily rule-based to a more goal-based system and the experience of the earlier Millennium Development Goals. Third, we review the negotiating history of the Sustainable Development Goals. Fourth, we elaborate on how the chapters are organized to address the three questions that guide the volume.

Goal Setting as a Global Governance Strategy

Governments and other political actors adopt goals at a global level to identify and publicize collective ambitions or aspirations in order to achieve some set of objectives, or at least to commit themselves publically to pursuing those objectives. By embracing international goals—through adopting such measures as declarations by conferences, summits, or the UN General Assembly—governments signal their interest in achieving such goals and possibly being held accountable for doing so. In return, goals are often expected to include measurable targets and time frames that are used in tracking progress. As a strategy of global governance, chapter 2 in this volume elaborates on these features, highlighting that goal setting aims to establish priorities that help combat the tendency for short-termism that would draw attention away from longer-term objectives.

Yet goal setting remains a contested governance strategy. Analysts are divided on its utility and effectiveness. Many international lawyers support the use of aspirational norms against which states can be held morally accountable. Others look at their value in terms of providing the foundations for formal institutional mechanisms to promote their diffusion and to sanction violators. Yet political "realists" tend to dismiss the setting of goals as a veneer for failures to achieve meaningful binding multilateral agreements. As Underdal and Kim (this volume, chapter 10) note, the adoption of seventeen Sustainable Development Goals as a package, along with the even wider 2030 Agenda for Sustainable Development, provides "scant guidance for prioritizing scarce resources," and there are no hierarchical governance arrangements internationally to ensure compliance. Still, they, as well as a number of other authors in this volume (for example, Bernstein, chapter 9; Voituriez, chapter 11; Gupta and Nilsson, chapter 12), though with varying degrees of caution, highlight the specific institutional and resource-mobilization efforts—some already emerging—to concretize implementation at multiple levels.

To some extent, the dichotomy between goals as standalone aspirations and goals as the foundations for longer-term commitments and meaningful action is false. Some goals initially embraced on their own terms have later had institutional structures attached (Szasz 1992). For example, the initial common goals expressed in the brief Atlantic Charter were later supplemented with formal institutional instruments at Dumbarton Oaks that created the United Nations. The pursuit of international human rights follows a similar evolutionary logic as states are increasingly caught up in a dense network of nongovernmental organizations and international institutions that are monitoring and advocating for stronger compliance (Simmons 2009; Hafner-Burton 2013; Sikkink 2011).

Broadly speaking, there are three types of international goal setting. First, some goals are solely aspirational. They may be presented by a small number of states hoping to catalyze longer-term support, or they may reflect a more general consensus regarding common aspirations for which governments may be held accountable. Examples include slavery prevention in the nineteenth century, human rights (Sandholtz and Stiles 2009), or the so-called "20/20" bargain put on the table for the 1992 UN Conference on Environment and Development in Rio de Janeiro (Speth 1992), which suggested that the developing countries reduce their greenhouse gas emissions by 20% and the industrialized countries increase foreign aid by 20%. Ultimately, aspirational goals may have unilateral effects, as governments choose to comply for reasons of belief. The ambition to limit global warming by 2°C above preindustrial levels is an example of such an aspirational goal. It was inscribed first into an EU agreement, then in a declaration of the Group of Eight major economies, and finally in the 2009 Copenhagen Accord of the parties to the 1992 UN Framework Convention on Climate Change. It made more concrete the abstract objective of the climate convention embedded in its article 2—"stabilization of greenhouse gas concentrations in the atmosphere at a level that would prevent dangerous anthropogenic interference with the climate system"—in the form of a numerical target.

A second type of goal setting consists of goals that start as aspirational but later acquire consensus and support through formal institutions that become attached to them for their enforcement and institutionalization. Once such goals are established, efforts to attain them proceed in a campaign mode and associated institutional development normally follows (Young, this volume, chapter 2). The Millennium Development Goals are an example. While originally devised as aspirational goals, the UN Secretariat subsequently devised a set of metrics to measure their achievement.

Other examples can be widely drawn from international environmental law, where broad aspirations are laid out in initial conventions that are then followed by more specific and enforceable protocols. Even without consensus on specific commitments, multilateral treaties, as Young (this volume, chapter 2) points out, may introduce specific regulatory mechanisms to operationalize goals, such as procedures to identify species at risk or levels for sustainable yields. This type of goal can shed light on issues that would otherwise be neglected.

A third type consists of goals to which (often novel) institutions and agencies are immediately attached. Principled consensus is here often broad and deep enough that governments create the institutional mechanisms for their immediate pursuit. Examples include the Bretton Woods institutions, but also the UN Environment Programme, created after the 1972 UN Conference on the Human Environment; the Commission on Sustainable Development, created to follow up on Agenda 21 agreed to at the 1992 UN Conference on Environment and Development; and the more recent High-level Political Forum on Sustainable Development, which will now follow up on the Sustainable Development Goals. However, in the latter case, the High-level Political Forum on Sustainable Development was created prior to deep consensus forming around the Sustainable Development Goals. Quite often, these types of goals do not lead to numerical targets but stay as broadly defined overarching goals, and institutional arrangements vary considerably in their means and capacities to follow up or institutionalize them.

The Sustainable Development Goals express some characteristics of each variety, but tend toward the first two, since the High-level Political Forum is not explicitly an implementing body and has so far little (or untested) authority and resources to directly support the goals, which will instead require buy-in, political action, and resource mobilization by a wide number of other actors and intermediary institutions at multiple levels (see part III of this volume). A proposal for a sustainability *Grundnorm* ("basic norm," see Young et al., this volume, chapter 3) might provide an opportunity for creating normative consensus; and the 2030 Agenda for Sustainable Development (UNGA 2015) may be tactically utilized to create such an opportunity according to the third type of goal setting.

The Context of the Sustainable Development Goals

Even though the Sustainable Development Goals arose in an overtly political context to replace the earlier Millennium Development Goals, they must be seen also as the latest instalment in an almost 30-year evolution of global governance that began with the popularization of the sustainable development concept. In this section, we now explore that conceptual and historical context.

Toward Sustainable Development as a Normative Goal

An especially important feature of this evolution is the gradual movement away from traditional governance mechanisms of norm promotion and rule making and toward goal setting, among other innovative governance mechanisms. While the reasons for this shift are varied, the general trend in global governance is well documented (Pauwelyn, Wessel, and Wouters 2014). The move toward innovative, multi-stakeholder, and goal-setting forms of global governance is especially discernable around sustainability concerns, as governments and stakeholders increasingly have sought new approaches given perceived limits, complexities, and failures of traditional global rule making (Kanie et al. 2013).

In addition, what started out as separate environment and development agendas has evolved over time toward much greater recognition of the interdependence of environmental, social, and economic systems. The World Commission on Environment and Development (known as the "Brundtland Commission") in 1987 articulated the first popular vision of sustainable development, which it defined as "development that meets the needs of the present without compromising the ability of future generations to meet their own needs" (World Commission on Environment and Development 1987). This definition has been used for decades as a reference point for the concept, even as it continues to prove challenging to measure given ambiguity in how to interpret it when applied to concrete policy. Still, the concept succeeded in not only cementing the importance of considering economic, social, and environmental dimensions of sustainable development as interdependent, but also by adding the time dimension to development through consideration of intergenerational equity, rather than focusing only on human well-being in a single generation.

The 1992 UN Conference on Environment and Development in Rio de Janeiro created further political momentum for action on sustainable development issues. However, it produced a particular interpretation of sustainable development consistent with the contemporary political and economic context. It focused attention almost exclusively on the environmental and development dimensions of the concept, within an overall liberal economic order. In practice, this interpretation prioritized economic growth and viewed market norms and mechanisms as the best way to simultaneously achieve environmental protection and development concerns (Bernstein 2001). Concretely, governments signed two major multilateral treaties at Rio—on climate change and biodiversity—as well as agreeing to the Rio Declaration, a statement of principles to guide action on environment and development, and Agenda 21, a detailed plan of action on a wide range of sustainable development issues. The Commission on Sustainable Development was established to follow up on the commitments made at Rio de Janeiro, specifically in Agenda 21.

Ten years later, the 2002 World Summit on Sustainable Development in Johannesburg assessed the state of implementation of Agenda 21 and called for further actions in the Johannesburg Plan of Implementation, but negotiated no new treaties. Instead, it promoted multisectoral public-private partnerships—so-called "type II outcomes"—as the primary means of implementation. Evaluations suggest that such partnerships have had, at best, mixed success. Many suffered from a lack of clear quantifiable goals and institutionalized monitoring, review, or evaluation mechanisms; significant underrepresentation of marginalized groups such as women, indigenous peoples, youth and children, and farmers; and relatively few partnerships actually geared toward implementing intergovernmental commitments (Biermann et al. 2007; Bäckstrand et al. 2012, 133–141; Pattberg et al. 2012; Bäckstrand and Kylsäter 2014). Around the time of the 2002 World Summit on Sustainable Development, the concept of sustainable development promoted within the United Nations also gradually moved to more self-consciously include three "pillars": environmental, economic, and social.

The 2012 UN Conference on Sustainable Development similarly did not include any negotiations on rules, but widened its focus from partnerships to include a variety of innovative governance and implementation mechanisms that involved mixes of government, stakeholder, foundation-based, and corporate participation and commitments. It also brought into greater focus than earlier summits the social dimension of sustainable development and emphasized the importance of integrating the three dimensions. Doing so acknowledged the reality of an increasingly fragmented and complex system of governance around the wide-ranging sustainable development agenda in which the United Nations was only one among many focal points. Thus the main means of implementation that the 2012 UN Conference on Sustainable Development recognized were some 730 voluntary commitments during the summit, and more than 700 more made by governments, international organizations, partnerships, action networks, and nonstate actors.

The Millennium Development Goals as Precursor

Broadly around the 2002 Johannesburg Summit, governments agreed also on the Millennium Development Goals, which are widely seen as one precursor to the current Sustainable Development Goals. The Millennium Development Goals were the result of a process that started in the 1990s, originally aiming at making development assistance more effective. At that time, international goals on development were agreed on in a number of conferences by the UN and the Organisation for Economic Co-operation and Development, some of which were eventually consolidated in the list of eight Millennium Development Goals, with originally 18 targets and 48 indicators, published in September 2001 in an annex to a "road map" produced by the UN Secretary-General. This road map stood in the broader context of the 2000 UN Millennium Declaration, which had already incorporated a number of specific targets (Manning 2010; Jabbour et al. 2012; Loewe 2012). The Millennium Development Goals were meant to guide global and national policies in the period toward 2015. In 2005, the list was expanded, with eventually 21 targets and 60 indicators, based on the work of an interagency and expert group (Manning 2010).

The Millennium Development Goals were significantly more limited than the new Sustainable Development Goals. They covered only a part of the sustainable development agenda, namely to eradicate extreme poverty and hunger; to achieve universal primary education; to promote gender equality and empower women; to reduce child mortality; to improve maternal health; to combat HIV/AIDS, malaria, and other diseases; to ensure environmental sustainability; and this all by developing a global partnership for development. Environmental concerns and questions of planetary stability—now much more central in the Sustainable Development Goals—were addressed merely in the seventh goal. This goal was specified in four targets on reversing natural resource degradation, reducing biodiversity loss, increasing access to safe drinking water and sanitation, and improving the lives of slum dwellers.

Unlike the current Sustainable Development Goals, the Millennium Development Goals essentially addressed developing countries only, with industrialized countries being involved mainly as funders of multilateral and national development agencies (addressed thus only in Millennium Development Goal 8, the global partnership for development). Also, the

Millennium Development Goals were not based on a widely carried, formal decision by the UN General Assembly, but developed rather by the UN Secretariat in the context of the Millennium Summit, even though drawing on previous intergovernmental conferences, consultations within and beyond the UN system, and inputs from governments.

There are both positive and negative lessons the experience of the Millennium Development Goals can offer. On the positive side, the Millennium Development Goals successfully mobilized support and brought attention to important but otherwise neglected global issues and communicated them in a concise and easily understandable way. Improvements related to the Millennium Development Goals include significantly reducing levels of extreme poverty, gender disparity in primary education, and gender inequality more generally. Other improvements included reductions in malaria-related diseases, improved access to clean drinking water, and mobilization of financial resources consistent with Millennium Development Goal 8, the "global partnership for development."

Nonetheless, the Millennium Development Goals have faced a number of criticisms. Chapters 6–8 of this volume evaluate some of these critiques. Some criticisms include gaps in levels of achievement among goals and among regions. They also failed to clearly articulate linkages between global goals and national or local goals and priorities. Part of the reason is that, by design, the UN Secretariat set the Millennium Development Goals at the global level, which had the effect of focusing attention on aggregate measures of progress. These aggregate measures did not necessarily help direct attention or resources to specific needs and demands at the national or local levels (Sumner 2009; Shepherd 2008; Browne 2014). Paradoxically, the ability to measure the success of the Millennium Development Goals against numerical benchmarks may have inhibited the ability of the Sustainable Development Goals to do the same. Indeed, the final form of the Sustainable Development Goals reflects repeated concerns raised in negotiations around a "one-size-fits-all" approach by frequently emphasizing the importance of country ownership, disaggregated data and measurement, consideration of different national and local capabilities and circumstances, and encouragement to formulate targets at the national level as well as leaving possibilities to create supplemental indicators at the national level.

Another set of criticisms concerns the lack of inclusiveness of the Millennium Development Goals. They mainly focused on three broad sets of issues from the Millennium Declaration: "development and poverty eradication," "protecting our common environment," and "meeting the special needs of Africa." Formulating them as a simple, memorable, and concise set of goals inevitably left other issues out. As Fukuda-Parr (2014) points out, the Millennium Development Goals encountered "unintended consequences" in diverting attention from other important issues and objectives. Still other critiques go to the nature of the targets. As the Millennium Development Goals were formulated around the idea of results-based management, issues such as human rights, equality, and governance effectiveness, where measuring progress is difficult or controversial, were not included (Alston 2005; Hulme 2007; Nelson 2007; Vandemoortele and Delamonica 2010; Browne 2014). Even in the case of included targets, purported causal connections between the Millennium Development Goals and measures of progress turned out to be questionable. For example, some argue that many ostensible achievements, especially on economic and poverty targets, owe substantially more to the economic boom in emerging economies during the period covered by the Millennium Development Goals, especially in China (Andresen and Iguchi, this volume, chapter 7).

Integrating Economic, Social, and Environmental Policies

Despite the fact that the new Sustainable Development Goals ostensibly replace the Millennium Development Goals—and as such explicitly incorporate and continue the pursuit of their core aim of ending poverty—the Millennium Development Goals are not the starting point of our conceptual discussions in this book. In our view, the Sustainable Development Goals constitute a fundamentally different approach to global problems that recognizes the interdependence of human societies and socio-ecological systems (Young et al., this volume, chapter 3). The *purpose* of the Sustainable Development Goals is to *capture the interconnections between issues*; that is, they encourage integrative and systemic approaches to global problems.

This is a vital difference. There is growing evidence that the earth system has entered a new epoch—the Anthropocene—in which humans now essentially shape planetary systems (Young et al., this volume, chapter 3). Humanity has become a systematic influence on natural systems, and human systems cannot be meaningfully disentangled from the natural ones on which they rely for vital resources. Given this historical transition and systemic transformation, chapter 3 engages the possibility of developing and institutionalizing a *Grundnorm* ("basic norm") of sustainability to underpin the Sustainable Development Goals. It would rest on some notion of respecting *planetary boundaries*, while also recognizing "the right of all people to improved well-being."

At the same time, the Sustainable Development Goals reflect a political outcome. As mentioned before, and as noted by a number of other chapters, the sustainable development concept itself reflects creative ambiguity, even as the attempt to integrate environmental, economic, and social goals reflects over 20 years of global negotiations and compromises since the 1992 Rio Summit. The Sustainable Development Goals explicitly claim to "integrate" and "balance" economic, social, and environmental purposes and to secure "interlinkages" among them, which raises questions about whether a coherent agenda will result, since including both modifiers in practice avoids difficult political debates about ultimate foundations. For example, as Bernstein (this volume, chapter 9) notes, the Sustainable Development Goals call for both "sustained" and "sustainable" economic growth and employment in Goal 8, but avoid any mention of planetary boundaries. At the same time, attempts had been made to include the concept in negotiations over the "growth" goal (*Earth Negotiations Bulletin* 2014), and respective Sustainable Development Goals mention the importance of securing natural resources or integration in policy of different dimensions of sustainable development. For example, Goal 12.2 states, "By 2030, achieve the sustainable management and efficient use of natural resources," and Goal 17.14, referring to means of implementation, states that such means should "enhance policy coherence for sustainable development."

Nearly all chapters highlight the challenge of operationalizing integrative action across the goals in a systemic way. These challenges range from integrating cross-cutting concerns such as better governance into implementing arrangements at multiple levels (Biermann et al., chapter 4), to creating integrated and system-oriented assessments and measures appropriate for monitoring and evaluating progress on goal attainment (Pintér, Kok, and Almassy, chapter 5), to the differentiated challenges and opportunities of integrated approaches to problems where there is low causal and normative consensus, such as education and urban sustainability, higher consensus such as food or water security, or mixed consensus such as public health (Haas and Stevens, chapter 6; on water see Yamada, chapter 8 and Gupta and Nilsson, chapter 12; on health see Andresen and Iguchi, chapter 7, all this volume). Andresen and Iguchi also argue that underachievement of the Millennium Development Goals stems from a lack of "fit" or mismatch between the structure of problems and institutional solutions, and the especially weak performance of the Millennium Development Goals on the environment is a case in point. While Millennium Development Goal 7 recognized environmental concerns, as a whole the Millennium Development Goals treated the environment largely in isolation, failing to recognize interlinkages among social, economic, and environmental concerns. Apart from some improvement on sanitation targets, fish stocks have continued to decline, deforestation has continued at an alarming rate, and global emissions of greenhouse gases have continued to rise (UN 2013). The importance of an integrated approach has also been emphasized in the scientific literature (Biermann 2014; Griggs et al. 2014; Kanie et al. 2014; Stafford-Smith et al. 2016). This increasing recognition of how systems are coupled and the need for integrative policies also highlights changes in understanding of global problems from the time of the Millennium Development Goals.

In sum, the Sustainable Development Goals emerged in the context of increased recognition that progress so far has been insufficient, global interdependencies and complexities have been increasing, and that the magnitude of response needed to address these complex challenges will require transformative changes in human behavior and governance systems. The experience of the Millennium Development Goals provided a positive template to break the stalemate of implementation of sustainable development policies and the Sustainable Development Goals received widespread support, not only among states in the North and South but among a wide range of stakeholders. Focusing on the Sustainable Development Goals also avoided the need to overcome a wide range of divisions that had plagued multilateral negotiations in a number of forums, which prevented binding commitments and progress on a range of issues, from trade to climate change (see, for example, Hale, Held, and Young 2013; Bernstein 2013). At the same time, bringing sustainable development into the context of the Millennium Development Goals and the mainstream development agenda (focused especially on poverty eradication) may now provide novel opportunities to substantively integrate environment and development, after 40 years of efforts. Indeed, the new Sustainable Development Goals and their central place in the post-2015 development agenda arguably mark a shift toward a new understanding of international development, at least in the UN context, as part of a wider, universal sustainability agenda.

Negotiating the Sustainable Development Goals

Let us now briefly review the negotiations that led to the eventual adoption of the Sustainable Development Goals. The government of Colombia, supported by Guatemala and the United Arab Emirates, first presented a proposal to establish Sustainable Development Goals during the High-level Dialogue on the Institutional Framework for Sustainable Development, held July 19–21, 2011, in Solo, Indonesia, in the lead-up to the 2012 UN Conference on Sustainable Development. When significant interest in the idea was voiced at various forums during the preparatory process, an informal consultation took place in Bogota, Colombia, in November 2011, with representatives from 30 countries. They saw the 2012 UN Conference on Sustainable Development as a critical opportunity to agree on a political commitment to sustainable development, and the need for a concrete approach as a basis for commitments to help guarantee the implementation of the 1992 "Agenda 21" and the 2002 Johannesburg Plan of Implementation. They emphasized the importance of the goal-oriented framework as a way to make it easier for governments and institutions to work together to reach common objectives.

Seven months later, the Sustainable Development Goals had become a centerpiece of the final outcome document of the 2012 UN Conference on Sustainable Development, "The Future We Want." Seven paragraphs (par. 245–51, this volume, annex 2) had been dedicated to the Sustainable Development Goals, and in the eyes of many, the agreement on a process to develop universal Sustainable Development Goals was "one of the most important political decisions of the Conference, given its centrality in helping to define the post-2015 development agenda" (*Earth Negotiations Bulletin* 2012).

The outcome document mandated the Sustainable Development Goals to be: action-oriented; concise and easy to communicate; limited in number; aspirational; global in nature; and universally applicable to all countries, while taking into account different national realities, capacities, and levels of development and respecting national policies and priorities. "The Future We Want" also stated that the process to establish them should be "coordinated" and "coherent with" the process to develop the post-2015 development agenda.

The process of establishing the Sustainable Development Goals attracted the most attention of negotiators in concluding the agreement at the 2012 UN Conference on Sustainable Development (*Earth Negotiations Bulletin* 2012). Initially, governments were divided on a number of issues. The European Union, for one, advocated a science-based process. Many developing countries, however, being often underrepresented in global scientific assessment processes, pushed to involve government experts (*Earth Negotiations Bulletin* 2012). In the end, governments agreed on the compromise to establish "an inclusive and transparent intergovernmental process on the Sustainable Development Goals that was open to all stakeholders with a view to developing global sustainable development goals to be agreed by the UN General Assembly." An Open Working Group was established of 30 representatives, nominated by governments through the five UN regional groups with the aim of ensuring "fair, equitable, and balanced geographic representation." The Open Working Group was expected to be constituted by the sixty-seventh session of the UN General Assembly in 2012, but intergovernmental negotiations on the selection of the 30 representatives and on modalities of the first Open Working Group meeting took longer than expected. Finally, on January 22, 2013, the UN General Assembly decided on membership of the Open Working Group in its resolution 67/555. Six seats were to be held by single countries (Benin, Congo, Ghana, Hungary, Kenya, and Tanzania). Nine seats were to be shared by two countries of similar regions (Bahamas and Barbados; Belarus and Serbia; Brazil and Nicaragua; Bulgaria and Croatia; Colombia and Guatemala; Mexico and Peru; Montenegro and Slovenia; Poland and Romania; and Zambia and Zimbabwe). Fourteen seats would be shared by trios of countries (Argentina, Bolivia, and Ecuador; Australia, the Netherlands, and the United Kingdom; Bangladesh, the Republic of Korea, and Saudi Arabia; Bhutan, Thailand, and Vietnam; Canada, Israel, and the United States; Denmark, Ireland, and Norway; France, Germany, and Switzerland; Italy, Spain, and Turkey; China, Indonesia, and Kazakhstan; Cyprus, Singapore, and United Arab Emirates; Guyana, Haiti, and Trinidad and Tobago; India, Pakistan, and Sri Lanka; Iran, Japan, and Nepal; and Nauru, Palau, and Papua New Guinea). The remaining seat would be shared by four countries (Algeria, Egypt, Morocco, and Tunisia). In practice, only a few groups coordinated their positions among those sharing a seat when making interventions, and many countries spoke on their own behalf. This made the deliberations in practice a more truly "open" working group, with about 70 countries, indicating the broad interest in being directly engaged in the formulation of the Sustainable Development Goals as opposed to leaving it to the 30 formally appointed members. Furthermore, the format helped ease traditional North-South confrontations, at least until the very final stage of the negotiation, by relaxing the rather tight coalitions that are often seen in UN negotiations and by providing opportunities for individual countries to speak on their own behalf.

The first session of the Open Working Group took place in March 2013 at the UN headquarters in New York, and elected as co-chairs Macharia Kamau of Kenya and Csaba Kőrösi of Hungary. The first eight sessions were devoted to exchanging views and ideas on a variety of thematic issues, with invited scientists and experts providing input. The rather lengthy stock-taking phase offered negotiators multiple opportunities for learning, which helped the Sustainable Development Goals to draw on concepts that went beyond traditional diplomatic language. On February 21, 2014, the co-chairs presented a document with 19 "focus areas," consolidating the stock-taking discussion and providing the basis for the subsequent five-month negotiation phase.

During these negotiations, the total number of goals moved between 16 and 19 (Bauer, Dombrowsky, and Scholz 2014). Delegates tried on many occasions to reduce the number of goals, following their mandate to make them "concise and limited in number." Also, a number of UN-sponsored reports had suggested shorter lists. For example, the High-level Panel of Eminent Persons on the Post-2015 Development Agenda, established by UN Secretary-General Ban Ki-moon, suggested 12 goals (High-level Panel 2013), and a report in June 2013 from the Sustainable Development Solutions Network—another initiative of the UN Secretary-General—suggested 10 goals (Leadership Council of the Sustainable Development Solutions Network 2013). In the end, the Open Working Group agreed to forward 17 goals with 169 targets for consideration by the UN General Assembly (UN 2014a).

This outcome also reflected UN Secretary-General Ban Ki-moon's synthesis report, "The Road to Dignity by 2030: Ending Poverty, Transforming All Lives and Protecting the Planet" (UN 2014b), an important input into the negotiations that helped frame the scope of the eventual 2030 Agenda for Sustainable Development (UNGA 2015). Notably, it identified the outcome of the Open Working Group as "the main basis for the post-2015 intergovernmental process." In part because of the inclusive process used to develop the Sustainable Development Goals, and in part because the draft agreement preceded the intensive phase of negotiations on the wider post-2015 agenda, the Sustainable Development Goals have mostly remained intact and at the center of the 2030 Agenda for Sustainable Development, the outcome of the entire process (UNGA 2015). Indeed, despite some misgivings about "sustainable development" in the title, which some developing countries saw as a debatable shift in language from the originally framed post-2015 "development agenda," in the end both the universal focus and the more encompassing concept of sustainable development prevailed. It may be fair to say that because major "post-MDGs" processes ended in 2013 and the Open Working Group was the only major intergovernmental process to discuss the agenda after that, the post-2015 development agenda had come to be discussed in the framework of the Sustainable Development Goals in the course of 2013–2014. This timing also helped elevate "sustainable development" to the mainstream international "development" agenda.

About This Book

This history and status of the Sustainable Development Goals is the backdrop against which the questions that animate this volume arise. Never before have such a detailed and extensive list of goals and targets been designed to drive the global governance agenda. These goals also have features that make them particularly fruitful and challenging subjects of study. The Sustainable Development Goals are at once more specific than previous efforts *and* less geared to secondary rule making at the time of delivery. They generally take the form of broad goals, with measurable targets and observable indicators, as well as procedures to track progress. Yet failure to achieve a goal has no direct consequence for targeted actors. Rather, the goals aim to mobilize not only the primary targets, but a multitude of actors and sectors. They provide benchmarks for progress but do not create specific responsibilities, obligations, or associated compliance mechanisms to induce actors to change behavior.

The Sustainable Development Goals also do not follow the typical pattern of aspirational goals that later led to specific rules or regulations. Instead, the targets under the Sustainable Development Goals either reiterate existing rules (say, as articulated in an international treaty), with their own separate institutional homes and mechanisms, or reflect longer-term goals like poverty eradication that continue to be stated in aspirational terms with no explicit language to suggest that rule making should follow on means to achieve the agreed-upon targets. However, they also do not exclude possibilities for future rule making, for example in areas where there are no international treaties (such as sustainable consumption and production).

This general lack of expectation for rule making to follow from the Sustainable Development Goals does not necessarily limit their effectiveness. Yet it makes it all the more important to identify specific mechanisms and conditions under which goals will produce desired outputs and outcomes. These are questions of governance, which brings us back to the three guiding questions for the volume listed earlier.

Part I of this volume broadly tackles the first question, on what characterizes goal setting, when it is an appropriate strategy in global governance, and what are the characteristics of global governance through goals as compared to other approaches such as rule making or norm promotion. First, Young (this volume, chapter 2) launches the discussion by highlighting several differences between rule making and goal setting as governance strategies, but he also suggests ways the two might work together. Young's chapter concludes with some suggestions to enhance the effectiveness of the Sustainable Development Goals given the pitfalls of international goal setting, especially in the absence of a close connection to rules, and also how building such linkages might improve the prospects of the Sustainable Development Goals.

Then, Young and colleagues (this volume, chapter 3) shift the focus to the underlying conditions in the twenty-first century that define the purpose of the Sustainable Development Goals with an extensive discussion of the implications of differences with the earlier Millennium Development Goals. Perhaps most controversially, the chapter challenges us to take seriously the meaning, and normative implications, of a sustainability framing in the Sustainable Development Goals by introducing the idea of a sustainability *Grundnorm* ("basic norm"), while at the same time acknowledging that this is in tension with the politics that produced the Sustainable Development Goals.

Whereas governance is recognized as central to the achievement of the Sustainable Development Goals, how to incorporate governance as a goal in its own right as well as an enabler for the implementation of goals, both globally and at subglobal levels, remains a key challenge for implementation and follow-up. Chapter 4 (Biermann et al., this volume) highlights this importance of recognizing a multifaceted view of governance on both counts, making the case that progress on the Sustainable Development Goals requires attention to "equitable" and "effective" governance as well as the traditional UN focus on "good" governance. Notably, these wider governance concerns also resonate with previous UN declarations and the broader 2030 Agenda for Sustainable Development (Browne 2014; UN 2014b). The chapter explores the politics of each of these categories, the degree to which the Sustainable Development Goals incorporate them, and their importance for creating conditions under which the goals can succeed through being integrated into governance institutions at multiple levels.

The fundamental underpinning of any governance system made up of goals and targets is measurement, which is addressed in chapter 5 of this volume. There, Pintér, Kok, and Almassy argue that the challenges of measurement of integrated problems of sustainability the Sustainable

Development Goals claim to embody require significantly rethinking the technical approach to monitoring and reporting on indicators that characterized earlier efforts, including the Millennium Development Goals. They thus propose a reform agenda that explicitly considers how the formulation of measures and indicators and use of data interact with the politics of transition or transformation underpinning the Sustainable Development Goals.

Part II of the volume then shifts the focus by looking back on what lessons can be drawn from earlier efforts to use goals as a governance strategy, including the Millennium Development Goals. Rather than simply offering an assessment of progress on the Millennium Development Goals, these chapters—guided especially by our second framing question—pay particular attention to previous efforts in areas the Sustainable Development Goals now recognize as sustainability challenges, with an eye to the prospects for the new goals to achieve their more demanding agenda.

Chapter 6 (Haas and Stevens, this volume) most strikingly highlights the so-far mixed record of previous efforts in sustainability governance, arguing that political and normative consensus, or lack thereof, is a major determinant of whether particular sustainability goals are likely to produce significant action. Their chapter points to the importance of social learning in order to generate consensual knowledge needed for transformation toward sustainability, but cautions that enabling conditions, including mechanisms to bring science to bear on policy as well as normative commitments to issue linkage, may be preconditions for such learning.

Chapters 7 and 8 evaluate lessons learned specifically from the Millennium Development Goals. Andresen and Iguchi (this volume, chapter 7) begin by highlighting an important problem in drawing any lessons from the Millennium Development Goals: the challenge of drawing causal connections between goals and outcomes given the wide range of possible external factors and lack of causal theory underlying the achievement of such goals. To address these limits of evaluating causality, they focus on Millennium Development Goal 4 and its targets related to the reduction of child mortality. They focus specifically on the Global Alliance for Vaccines and Immunization, a campaign mobilized explicitly in response to this goal, and the role of Norway in playing a leadership role on health-related Millennium Development Goals, in order to assess the importance of national leadership for goal attainment.

Yamada (this volume, chapter 8) shifts the focus to the corporate sector. The role of this sector as both a target of and a potential partner in addressing sustainability is increasingly recognized. It is seen by many as central in the effort to make progress on the Sustainable Development Goals. Yamada looks specifically at the Global Compact's CEO Water Mandate in addressing water security as part of Millennium Development Goal 7, the only UN-sponsored multi-stakeholder forum explicitly designed for this purpose. Given that water governance lacks strong international rules for states or corporations, Yamada identifies several strategies, such as *activation* (mobilizing corporations, such as through a stakeholder forum), *orchestration* (in this case Millennium Development Goal 7 vis-à-vis its stakeholders), and *modulation* (by creating an incentive structure for corporations to commit to, in this case, water stewardship).

Part III of the book moves to examine the conceptual and practical challenges of governing through goals in promoting sustainability and the prospects for achieving such a demanding new agenda—the fundamental political challenge posed by our volume's third animating question.

Bernstein (this volume, chapter 9) starts the discussion by arguing that even if the Sustainable Development Goals were perfectly designed, they would still require appropriate governance arrangements to diffuse them and integrate them into institutions and practices, both throughout the system and at regional, national, and local levels. The institutional and political challenges will be formidable, especially given the integrative nature of the Sustainable Development Goals compared to previous goal-setting exercises. Governance through this newer type of goal needs to build a broad diverse coalition by mobilizing a wide range of public and private actors, and provide incentives to change behavior in pursuit of the objectives identified by the goals. This is especially necessary for the Sustainable Development Goals, given their overarching purpose to steer, and ultimately transform, societies toward sustainability. This challenge requires us to move from a vision of governance as implementation to viewing it as a necessary means and enabler of catalyzing and releasing the potential of relevant implementers and stakeholders. The centrality of governance—recognized in Goal 16 through its focus on justice, inclusion, and accountable institutions—also raises questions about whether the Sustainable Development Goals as a whole fully capture what is needed for necessary action and mobilization (Biermann et al., this volume, chapter 4).

Monitoring and measuring thus become central mechanisms of governance, creating a host of practical and political dilemmas for countries at different levels of development and different domestic circumstances (Pintér, Kok, and Almassy, this volume, chapter 5). While the Sustainable Development Goals explicitly aim to "take into account" national circumstances, problems of capacity, analysis, and political sensitivities complicate this requirement. For example, many countries lack statistical capacity, disaggregated data, or monitoring capacity. Even with excellent data collection, comparable measurement and systemic or integrated analysis of data pose significant challenges for the Sustainable Development Goals.

These challenges raise inherently political issues and potential conflicts that can generally only be resolved in follow-up through institutional arrangements that may also serve to orchestrate or create linkages among existing agreements. This is the main focus of Underdal and Kim's analysis (this volume, chapter 10). As they argue, concrete targets can lead to either unproductive strategic behavior or encoding; that is, lowest-common-denominator targets as opposed to raising aspirations even for countries that are already doing better or are willing to make stronger commitments. Similarly, coordination or "orchestrating" from existing understandings and initiatives can be difficult, especially when goals and targets reflect particular interests traditionally put forward by issue-specific institutions at international and national levels. The chapters in this volume by Haas and Stevens (chapter 6), Yamada (chapter 8), and Underdal and Kim (chapter 10) explicitly address these challenges at the intersection of political conflicts and consensual knowledge, as well as the possibilities of normative underpinnings required to create necessary linkages.

Importantly, because global goals establish priorities for allocating attention and scarce resources among competing objectives, mobilizing such attention and support—including financial, technical, and human resources—becomes a central purpose of governance. Voituriez and colleagues (this volume, chapter 11) thus examine in detail the challenge of financing and the implications of the 2015 Financing for Development conference in Addis Ababa (UN 2015). In addition, this volume studies practical questions of leadership, coherence, and resource mobilization—in other words, of actually creating working governance systems around specific Sustainable Development Goals or for action on the goals as a whole (Biermann et al., chapter 4; Andresen and Iguchi, chapter 7; Bernstein, chapter 9; Voituriez et al., chapter 11; Gupta and Nilsson, chapter 12, all this volume).

Finally, promoting and facilitating coherent and integrated governance systems for action on the Sustainable Development Goals at the country level is especially challenging. Gupta and Nilsson analyze, in a richly detailed case study with a focus on water, the challenge of implementing goals at the national level (this volume, chapter 12). They make a strong case for the need to integrate governance across scales, which requires the mobilization of multiple actors and resources at multiple scales. Doing so also raises important questions of accountability and coherence given limits of strong centralized control or accountability mechanisms—and questions about their desirability—as well as existing power and resource differentials among and between implementing and affected actors and communities. As the Sustainable Development Goals enter an already crowded institutional environment at both international and national levels, coherence becomes especially challenging.

Conclusion

Ultimately, this volume aims to better understand the prospect for improved global problem solving through goal setting, with a view to broader societal and governance transformations. Do the Sustainable Development Goals similarly offer a promise of hope from the perspective of a hungry rural farmer in Mali, an unemployed inner-city Detroit machinist, a struggling Chinese laborer, a resident of Tuvalu who worries about whether her home on the island will still be inhabitable when her children are grown, or a Pakistani villager with little access to potable water? In the absence of international sanctioning mechanisms and significant financial transfers or other major forms of directed resource mobilization, strong state-level commitments are initially unlikely. Yet the Sustainable Development Goals may be an important step in the longer-term development of more widely shared norms of sustainability around which states can craft policies, actors can mobilize, and institutional mechanisms can adapt and support. This volume seeks to clarify the processes of goal setting and their implications as well as the prospects for moving from goal setting to meaningful action.

A goal-oriented approach to sustainability issues emerged at an opportune time. Based on the experience of the Millennium Development Goals, the Sustainable Development Goals could serve to raise the level of ambition and reduce the gap between the prevailing political pragmatism, on the one hand, and what many scientists argue is needed to secure a safe operating space for the earth's life-support systems, on the other hand. However, the experience of the Millennium Development Goals also provides a cautionary tale, where it is difficult to point to any specific rule making or institutionalized implementation mechanisms that followed from the Millennium Development Goals, even as they arguably succeeded in mobilizing a range of actors and resources and achieved some of their goals.

Goal-oriented approaches can also produce unintended effects. Because they mobilize support and attention, goals can also distort priorities "by displacing attention from other objectives, disrupting ongoing initiatives and alliances, creating perverse incentives, and undermining alternative policy analysis" (Fukuda-Parr, Yamin, and Greenstein 2014; also Underdal and Kim, this volume, chapter 10). By focusing both on the specific processes and governance arrangements around the Sustainable Development Goals, and broader questions about goal setting as a governance strategy, all chapters in this volume aim to shed light on these opportunities and ways to anticipate and minimize the risks. The research and findings in these chapters thus not only offer a detailed analysis of governance of and for the Sustainable Development Goals, but also the first scholarly treatment of this increasingly prominent and novel form of global governance through goals.

References

Alston, Philip. 2005. Ships Passing in the Night: The Current State of the Human Rights and Development Debate Seen through the Lens of the Millennium Development Goals. *Human Rights Quarterly* 27 (3): 755–829.

Bäckstrand, Karin, Sabine Campe, Sander Chan, Ayşem Mert, and Marco Schäferhoff. 2012. Transnational Public-Private Partnerships. In *Global Environmental Governance Reconsidered*, ed. Frank Biermann and Philipp Pattberg, 123–147. Cambridge, MA: MIT Press.

Bäckstrand, Karin, and Mikael Kylsäter. 2014. Old Wine in New Bottles? The Legitimation and Delegitimation of UN Public-Private Partnerships for Sustainable Development from the Johannesburg Summit to the Rio+20 Summit. *Globalizations* 11 (3): 331–347.

Bauer, Steffen, Ines Dombrowsky, and Imme Scholz. 2014. *Post 2015: Enter the UN General Assembly. Harnessing Sustainable Development Goals for an Ambitious Global Development Agenda.* Bonn: German Development Institute.

Bernstein, Steven. 2001. *The Compromise of Liberal Environmentalism.* New York: Columbia University Press.

Bernstein, Steven. 2013. Rio+20: Sustainable Development in a Time of Multilateral Decline. *Global Environmental Politics* 13 (4): 12–21.

Biermann, Frank. 2014. *Earth System Governance: World Politics in the Anthropocene.* Cambridge, MA: MIT Press.

Biermann, Frank, Man-san Chan, Ayşem Mert, and Philipp Pattberg. 2007. Multi-stakeholder Partnerships for Sustainable Development: Does the Promise Hold? In *Partnerships, Governance and Sustainable Development: Reflections on Theory and Practice*, ed. Pieter Glasbergen, Frank Biermann and Arthur P. J. Mol, 239–260. Cheltenham: Edward Elgar.

Browne, Stephen. 2014. A Changing World: Is the UN Development System Ready? *Third World Quarterly* 35 (10): 1845–1859.

Earth Negotiations Bulletin. 2012. Summary of the United Nations Conference on Sustainable Development: 13–22 June 2012. *Earth Negotiations Bulletin* 27 (51).

Earth Negotiations Bulletin. 2014. Summary of the Second Meeting of the High-level Political Forum on Sustainable Development: 30 June–9 July 2014. *Earth Negotiations Bulletin* 33 (9).

Fukuda-Parr, Sakiko. 2014. Global Goals as a Policy Tool: Intended and Unintended Consequences. *Journal of Human Development and Capabilities* 15 (2–3): 118–131.

Fukuda-Parr, Sakiko, Alicia Ely Yamin, and Joshua Greenstein. 2014. The Power of Numbers: A Critical Review of Millennium Development Goal Targets for Human Development and Human Rights. *Journal of Human Development and Capabilities* 15 (2–3): 105–117.

Griggs, David, Mark Stafford Smith, Johan Rockström, Marcus C. Öhman, Owen Gaffney, Gisbert Glaser, Norichika Kanie, Ian Noble, Will Steffen, and Priya Shyamsundar. 2014. An Integrated Framework for Sustainable Development Goals. *Ecology and Society* 19 (4): 49.

Hafner-Burton, Emilie. 2013. *Making Human Rights a Reality*. Princeton: Princeton University Press.

Hale, Thomas, David Held, and Kevin Young. 2013. *Gridlock: Why Global Cooperation is Failing When We Need it Most*. Cambridge: Polity Press.

High-level Panel, High-level Panel of Eminent Persons on the Post-2015 Development Agenda. 2013. *A New Global Partnership: Eradicate Poverty and Transform Economies through Sustainable Development*. New York: United Nations.

Hulme, David. 2007. The Making of the Millennium Human Development Meets Results-based Management in an Imperfect World. Brooks World Poverty Institute Working Paper 16: 1–26.

Jabbour, Jason, Fatoumata Keita-Ouane, Carol Hunsberger, Roberto Sánchez-Rodríguez, Peter Gilruth, Neeyati Patel, Ashbindu Singh, et al. 2012. Internationally Agreed Environmental Goals: A Critical Evaluation of Progress. *Environmental Development* 3: 5–24.

Kanie, Norichika, Naoya Abe, Masahiko Iguchi, Jue Yang, Ngeta Kabiri, Yuto Kitamura, Shunsuke Managi, et al. 2014. Integration and Diffusion in Sustainable Development Goals: Learning from the Past, Looking into the Future. *Sustainability* 6 (4): 1761–1775.

Kanie, Norichika, Peter M. Haas, Steinar Andresen, Graeme Auld, Benjamin Cashore, Pamela S. Chasek, Jose A. Puppim de Oliveira, et al. 2013. Green Pluralism: Lessons for Improved Environmental Governance in the 21st Century. *Environment: Science and Policy for Sustainable Development* 55 (5): 14–30.

Leadership Council of the Sustainable Development Solutions Network. 2013. An Action Agenda for Sustainable Development: Report for the UN Secretary-General.

Loewe, Markus. 2012. *Post 2015: How to Reconcile the Millennium Development Goals (MDGs) and the Sustainable Development Goals (SDGs)?* Bonn: German Development Institute.

Manning, Richard. 2010. The Impact and Design of the MDGs: Some Reflections. *IDS Bulletin* 41 (1): 7–14.

McArthur, John W. 2014. The Origins of the Millennium Development Goals. *SAIS Review (Paul H. Nitze School of Advanced International Studies)* 34 (2): 5–24.

Nelson, Paul J. 2007. Human Rights, the Millennium Development Goals, and the Future of Development Cooperation. *World Development* 35 (12): 2041–2055.

Pattberg, Philipp, Frank Biermann, Sander Chan, and Ayşem Mert, eds. 2012. *Public-Private Partnerships for Sustainable Development: Emergence, Influence, and Legitimacy*. Cheltenham: Edward Elgar.

Pauwelyn, Joost, Ramses A. Wessel, and Jan Wouters. 2014. When Structures Become Shackles: Stagnation and Dynamics in International Lawmaking. *European Journal of International Law* 25: 733–763.

Ruggie, John G. 1996. *Winning the Peace*. Columbia University Press.

Sandholtz, Wayne, and Kendall Stiles. 2009. *International Norms and Cycles of Change*. Oxford: Oxford University Press.

Shepherd, Andrew. 2008. *Achieving the MDGs: The Fundamentals. ODI Briefing Paper 43*. London: Overseas Development Institute.

Sikkink, Kathryn. 2011. *The Justice Cascade: How Human Rights Prosecutions Are Changing World Politics*. New York: Norton.

Simmons, Beth A. 2009. *Mobilizing for Human Rights. International Law in Domestic Politics*. Cambridge: Cambridge University Press.

Speth, Gustave. 1992. A Post-Rio Compact. *Foreign Policy* 88: 145–161.

Stafford-Smith, Mark, David Griggs, Owen Gaffney, Farooq Ullah, Belinda Reyers, Norichika Kanie, Bjorn Stigson, Paul Shrivastava, Melissa Leach, and Deborah O'Connell. 2016. Integration: The Key to Implementing the Sustainable Development Goals. *Sustainability Science*. DOI:10.007/s11625-016-0383-3.

Sumner, Andy. 2009. Rethinking Development Policy: Beyond 2015. *Broker* 14: 8–13.

Szasz, Paul C. 1992. International Norm-making. In *Environmental Change and International Law: New Challenges and Dimensions*, ed. Edith Brown Weiss, 41–80. Tokyo: UN University Press.

UN, United Nations. 2013. Millennium Development Goals Report. New York: United Nations.

UN, United Nations. 2014a. Report of the Open Working Group of the General Assembly on Sustainable Development Goals. UN Doc. A/68/970.

UN, United Nations. 2014b. The Road to Dignity by 2030: Ending Poverty, Transforming All Lives and Protecting the Planet: Synthesis Report of the Secretary-General on the Post-2015 Sustainable Development Agenda. UN Doc. A/69/700.

UN, United Nations. 2015. Outcome document of the Third International Conference on Financing for Development: Addis Ababa Action Agenda. UN Doc. A/CONF.227/L.1.

UNGA, United Nations General Assembly. 2015. Transforming Our World: The 2030 Agenda for Sustainable Development. Draft resolution referred to the United Nations summit for the adoption of the post-2015 development agenda by the General Assembly at its sixty-ninth session. UN Doc. A/70/L.1.

Vandemoortele, Jan, and Enrique Delamonica. 2010. Taking the MDGs Beyond 2015: Hasten Slowly. *IDS Bulletin* 41 (1): 60–69.

Williams, Andrew. 1998. *Failed Imagination?* Manchester University Press.

World Commission on Environment and Development. 1987. *Our Common Future*. Oxford: Oxford University Press.

I Goal Setting as a Governance Strategy

2 Conceptualization: Goal Setting as a Strategy for Earth System Governance

Oran R. Young

The challenge of meeting needs for governance has emerged as a central concern in many social settings, not least in international society where there is no overarching government to take responsibility and where we are confronted with the formidable complexities of integrating the biophysical, economic, and social forces affecting the achievement of sustainable development on a global scale. In considering ways to meet this challenge, we tend to think first of regulatory arrangements, emphasizing the development of rules and focusing on issues relating to the implementation of the rules and procedures that are useful in eliciting compliance on the part of those subject to the rules (Chayes and Chayes 1995). If we think of governance in generic terms as a social function centered on steering individuals or groups toward desired outcomes, however, we can conceptualize goal setting and efforts to meet targets associated with key goals as a distinct strategy for fulfilling needs for governance.

In this chapter, I explore the nature of goal setting as a governance strategy, analyze conditions under which goal setting can prove effective as a steering mechanism, consider ways to enhance the effectiveness of goal setting in various settings, and comment on the relevance of this way of thinking to the effort under the auspices of the United Nations to develop a set of Sustainable Development Goals. My argument is intended to be empirical rather than normative or prescriptive in character. I make no effort to pass judgment on the relative merits of rule making and goal setting as distinct governance strategies. Rather, I seek to shed light on goal setting as a way of responding to needs for governance that has received far less attention than rule making among those who think about governance at the international or global level.

My purpose is not to contribute directly to the efforts of those working on the implementation of Sustainable Development Goals, and I make no attempt to produce specific recommendations. But what I have to say may prove useful to those seeking to maximize the effectiveness of the Sustainable Development Goals, helping them to avoid some common pitfalls of goal setting as a governance strategy, and to arrive at an interpretation of goals that are realistic in political terms and at the same time appropriately aspirational in normative terms.

I proceed as follows. The first substantive section explores the basic character of goal setting as a governance strategy and differentiates it from the more familiar idea of rule making. The second section comments on circumstances under which it may be expedient to join goal setting and rule making to form integrated governance systems. The third section then identifies a number of pitfalls that plague efforts to employ goal setting to good effect in large-scale settings like international society. In the fourth section, I turn to the issue of effectiveness and set forth some general observations regarding the determinants of the success or effectiveness of goal setting as a governance strategy. The fifth section discusses procedures or mechanisms that may prove helpful to those desiring to enhance the effectiveness of goal setting. In the final section, I offer some initial reflections on the implications of this train of thought for the development and implementation of the Sustainable Development Goals themselves.

Goal Setting as a Governance Strategy

Goal setting seeks to steer behavior by (i) establishing priorities to be used in allocating both attention and scarce resources among competing objectives, (ii) galvanizing the efforts of those assigned to work toward attaining the goals, (iii) identifying targets and providing yardsticks or benchmarks to be used in tracking progress toward achieving goals, and (iv) combating the tendency for short-term desires and impulses to distract the attention or resources of those assigned to the work of goal attainment. Goal setting thus differs from rule making, which seeks to guide the behavior of key actors by articulating rules (and associated regulations) and devising compliance mechanisms whose purpose is to induce actors to adjust their behavior accordingly.

To make these somewhat abstract ideas concrete, think of the capital campaigns that universities, hospitals, libraries, public radio stations, and various charitable organizations launch from time to time as a source of illustrative examples. The usual procedure is to set a concrete goal defined in monetary terms, lay out some plans regarding the use of the money to be mobilized, specify a target date for meeting the goal, publicize the campaign vigorously, and create a highly visible system for tracking progress.

The idea is to activate known supporters, while at the same time identifying and signing up new supporters. Such campaigns not only serve to focus the efforts of regular staff members and to mobilize enthusiastic volunteers; they also serve to mobilize resources and set the relevant organizations on paths that are likely to shape their programmatic development for years to come. The effectiveness of capital campaigns is by no means assured. There are highly paid experts whose business it is to advise organizations on when to launch a capital campaign and how to decide on an appropriate goal. But well-planned and well-timed campaigns regularly prove successful in meeting their goals. Some even exceed their goals, an occurrence that allows leaders to make credible claims regarding their prowess as fundraisers.

This illustration is useful as a means of revealing the defining features of goal setting as a steering mechanism. Taking this conception of goal-setting as a point of departure, we can identify relatively well-defined examples of the use of this governance strategy at the national and international levels. Prominent cases at the national level include the redirection of the US economy during 1942–1943 to fulfill the goal of turning the United States into an "arsenal of democracy" in the fight against the Axis powers and, perhaps even more clearly, the work of the US Apollo Project launched by the Kennedy Administration and designed to put a person on the moon by the end of the 1960s. Undoubtedly, the most prominent recent example of goal setting at the international level features the development and implementation of the Millennium Development Goals launched in 2000 through the adoption of the UN General Assembly's Millennium Declaration.

Although the particulars of goal setting vary greatly from one situation to another, all efforts to make use of goal setting as a governance strategy share three distinctive features. Goal setting requires, to begin with, an ability to establish well-defined priorities and to cast them in the language of explicit goals. The whole point of goal setting is to single out a small number (sometimes just one) of concerns and to accord them priority in the allocation of scarce resources, including staff time and political capital. Once the goals are established, efforts to attain goals normally proceed in campaign mode. The essential idea is to galvanize attention and mobilize resources to make a sustained push to achieve measurable results within a fixed time frame. The goal of the Apollo Project to put a person on the moon within a decade provides a striking but representative example. In addition, goal setting requires an effort to devise clear-cut metrics to track progress over time (see Pintér, Kok, and Almassy, this volume, chapter 5).

Millennium Development Goal 1, aimed at halving the number of people living on less than US $1 a day by 2015, provides a good example, though collecting the actual data needed to operationalize a metric of this sort may prove difficult under real-world conditions. Not only do tracking mechanisms make it possible to measure progress toward goal attainment; they are also useful as a means of urging all those involved to redouble their efforts to fulfill the goal by the predetermined cutoff date.

The essential premise of goal setting as a governance strategy consequently differs from the premise underlying rule making. Whereas rule making features the formulation of behavioral prescriptions (for example, requirements and prohibitions) and directs attention to matters of compliance and enforcement, goal setting features the articulation of aspirations and directs attention to procedures for generating enthusiasm among supporters and maximizing the dedication needed to sustain the effort required to reach more or less well-defined targets. Moreover, whereas goal setting normally features the mounting of a campaign designed to attain goals within a specified time frame, rule making features the articulation of behavioral prescriptions expected to remain in place indefinitely.

Linking Goal Setting and Rule Making

Both goal setting and rule making can and often do operate on their own as distinct strategies for addressing governance needs. Consider, for illustrative purposes, the contrast between the goal-setting strategy reflected in the earlier Millennium Development Goals and the rule-making approach embodied in many regimes aimed at achieving sustainable results in the harvesting of living or renewable resources. The Millennium Development Goals call on parties to mount a campaign to achieve goals like eradicating extreme poverty and hunger, reducing childhood mortality, and combating HIV/AIDS, malaria, and other diseases, all within a time frame extending to the end of 2015. Regimes governing the use of renewable resources, by contrast, typically rely on rules governing such things as permits or licenses required to engage in harvesting the resources, quotas, open and closed seasons, restrictions on gear types, the treatment of by-catches, and so forth. Often, there is little overlap regarding the mechanisms used in the two types of governance systems.

Nevertheless, the two strategies are not mutually exclusive. They may prove complementary in specific situations. Even more to the point, goal setting and rule making may become elements in integrated governance systems. Consider as examples the goal of avoiding "dangerous anthropogenic interference with the climate system" articulated in the UN Framework Convention on Climate Change as the overall objective of the climate regime (1992, art. 2) and the goal of achieving maximum sustainable yield or even optimum sustainable yield embedded in regimes for fisheries or marine mammals. A case in point is the 1946 International Convention on the Regulation of Whaling. In such cases, regulatory arrangements are introduced as means of steering behavior toward the achievement of explicit goals. Situations of this sort are common and deserve systematic attention in any comprehensive account of governance strategies. But the principal focus of this chapter is on goal setting as a distinct strategy of governance in situations featuring free-standing goals like the Millennium Development Goals, and now the Sustainable Development Goals.

Common Pitfalls of Goal Setting at the International Level

Goal setting is intuitively appealing, both because we all have experience with setting goals in our private lives and because most of us feel that the pursuit of such personal goals has played or can play an important role as a means of focusing and guiding our efforts—or, in other words, as a form of self-governance. But can we generalize from our personal experiences with goal setting employed as a strategy for achieving self-governance to the use of goal setting as a strategy for achieving collective governance at the level of international society?

In addressing this question, a number of differences between self-governance and collective governance come into focus that are apt to pose problems for the use of goal setting in international society. Consider the following issues as illustrations of this proposition.

Setting the goals. Setting priorities is difficult under the best of circumstances. Individuals often experience inner conflict in the effort to establish personal priorities. But when goal setting is a collective endeavor involving some form of negotiation or consensus-building among a relatively large number of self-interested actors (for example, states in international society), there is a danger that (i) the group will end up with too many goals to be helpful in establishing priorities and allocating resources; (ii) the goals chosen will be framed in vague terms that are hard to operationalize, much less to monitor; or (iii) individual goals included in a package will be incompatible or even contradictory.

Tracking progress. It is easy to track progress in the case of a fund-raising campaign where the goal is stated in monetary terms. Devices like a "barometer" showing the percentage of the total raised at any given time are familiar benchmarks used in this context. But many international goals are framed in terms that are hard to measure and result in the use of procedures that are inadequate or even misleading. In tracking progress toward improving human welfare, for example, there is a danger of overemphasizing easily measurable criteria (for example, GDP per capita), underemphasizing values that are critical but harder to measure (for example, well-being or human development), or ending up with a high level of uncertainty or disagreement about the issue of tracking progress toward goal attainment (for example, the operational meaning of a concept like social welfare; Karabell 2014).

Behavioral mechanisms. The incentives and pressures to stick with goals may be inadequate to steer behavior at the international level, especially when the going gets tough due to competing demands or shortages of resources during periods of economic or social stress. The mechanisms most likely to produce good results in pursuing goals may differ sharply from those that are effective in promoting compliance with rules and regulations. The success of goal setting, for example, generally requires building a coalition dedicated to making progress, whereas eliciting compliance from individual subjects is essential to the success of rule making. The effectiveness of goal setting as a governance strategy may differ as well to the extent that the strategy is used as a stand-alone tool for solving governance problems, as a supplement to a rule-based system, or as a way of identifying targets articulated in a rule-based system.

Opportunity costs and benefits. A focus on goal setting can deflect attention and divert resources from efforts to build effective governance systems through the creation and implementation of rule-based regimes on issues like climate change or the loss of biological diversity. In cases where resources are limited, it may be important to confront choices regarding the menu of goal setting and rule-making options. On the other hand, there may be opportunities to achieve synergy either by using these tools sequentially or employing them together.

Complacency. More generally, goal setting can breed a kind of complacency rooted in a sense that once appealing goals are established, there is no need to invest time and energy in more demanding forms of problem solving. This creates a danger that those who are unable or unwilling to address governance needs through regulatory measures will use goal setting as a diversion to deflect public attention from their unwillingness to take governance problems seriously.

Determinants of the Success of Goal Setting

The challenge of evaluating the effectiveness of goal setting has much in common with the parallel challenge in the case of rule making (Young 2011). To begin with, the familiar distinction among outputs, outcomes, and impacts applies just as well to the evaluation of success regarding goal setting as it does to the effectiveness of rule making. Outputs relating to goal setting encompass the articulation of targets and indicators associated with specific goals as well as the establishment of organizational arrangements to oversee the effort to attain goals. Outcomes refer to behavioral adjustments on the part of both states and nonstate actors intended to promote progress toward goal attainment. Impacts, then, involve progress toward fulfilling the goals themselves. As in the case of rule making, the causal chain becomes longer, more complex, and harder to pin down as we move from outputs to impacts. It is relatively easy to establish a causal connection between the articulation of goals and the establishment of organizational arrangements designed to promote their attainment. It is another matter to demonstrate such a connection between goal setting and actual progress toward fulfilling the relevant goals. Take the case of the Millennium Development Goal focusing on poverty eradication as an example in this regard. As a number of analysts have observed, there has been noticeable progress toward eradicating extreme poverty in the period since the adoption of the Millennium Declaration in 2000 (Sachs 2015). But can we show an unambiguous causal relationship between the adoption and implementation of the Millennium Development Goals and this development? At a minimum, this is a situation featuring causal complexity. The Millennium Development Goals may have played a role in reducing extreme poverty, but so have a variety of other factors having to do with economic growth, social change, and public policy in developing countries like China (see Haas and Stevens, this volume, chapter 6; Andresen and Iguchi, this volume, chapter 7).

That said, there is no reason to expect goal setting to perform equally well (or poorly) in addressing needs for governance under all conditions. What may work in a close-knit community whose members are dedicated to the good of the whole, for instance, may fall flat as a steering mechanism in a group whose members are fiercely independent actors who prioritize efforts on the part of individual members to make their own way under most conditions. This suggests the value of launching an inquiry into the determinants of success with regard to the use of goal setting as a governance strategy. Here, I set forth some preliminary observations about four important sets of conditions relating to: (i) the nature of the problem, (ii) the character of the actors, (iii) the principal features of the setting, and (iv) the mobilization of support in specific cases.

Nature of the problem. The problems that generate needs for governance are not all alike; some are likely to be more amenable to goal-setting strategies than others. In the case of lumpy collective goods, for instance, those seeking to supply governance must overcome not only the familiar challenge of the free-rider problem, but also complications arising from the fact that none of the good can be supplied until a specific (possibly high) threshold is reached. Where collective goods are continuous, by contrast, early efforts allowing for initial supplies of the goods may encourage members of the group to make additional contributions to obtain more. In cases where the problem is spatially defined (for example, protecting biodiversity in mega-diverse countries), there are often complex issues regarding the extent to which outsiders can or should contribute and the modalities to be used in channeling the contributions they do make. In cases where the problem is finite (for example, eradicating a disease), fulfilling a goal may solve the problem once and for all. Where the problem is continuous (for example, controlling emissions of greenhouse gases), on the other hand, goal setting may prove less effective as a governance strategy, since it may be impossible to make clear-cut assessments of goal attainment. Beyond this lie differences in the scale and scope of the problem. Achieving the maximum sustainable yield in a fishery that is spatially limited and lightly used by a few easily identifiable fishers is one thing; pursuing the same goal in a far-flung fishery heavily used by many fishers who engage in illegal, unregulated, and unreported activities is another. This discussion introduces what is known in research on international cooperation as the problem of fit (Young, King, and Schroeder 2008). In a word, success in goal attainment will depend on the extent to which the goals selected and the procedures introduced to pursue them are compatible with the defining features of the problem they seek to solve.

Character of the actors. Achieving success is likely to be determined in part by the extent to which the behavior of the actors reflects the logic of consequences versus the logic of appropriateness (March and Olsen 1998). Where the logic of consequences prevails, success will require an appeal to actors on the basis of calculations of benefits and costs. When the logic of appropriateness takes hold, by contrast, tying goal setting to norms and principles may prove effective. More generally, goal setting may work well in situations where the objective can be integrated into a coherent social narrative, so that it becomes part of how actors perceive their identity and organize their thinking about governance. In cases where contributing to the common good is culturally prescribed, for example, it may be comparatively easy to attain goals that involve the supply of collective or public goods. At the international level, there is the added challenge of dealing with actors that are themselves collective entities. Not only does this introduce the familiar issue of two-level games in thinking about the effectiveness of goal setting; it also makes it important to think about a host of issues arising from the facts that living up to international goals may prove highly contentious in the domestic politics of individual states and that commitments to such goals may wax and wane over time as governments come and go at the domestic level. A particularly important challenge arises when a new administration takes office seeking to differentiate itself from its predecessor and looking for ways to free up resources to provide the wherewithal to launch new policy initiatives. Forces of this sort undoubtedly play a role in the common failure of states to fulfill the pledges they make regarding financial aid to developing countries, or aid needed to attain goals like protecting the earth's climate system or reducing the loss of biological diversity.

Features of the setting. Goal attainment is commonly affected by features of the prevailing social setting, such as the number of actors involved, the extent to which the actors are linked together through ties of common interests or cultural affinity, the degree to which affluence eases the burden of making contributions toward the attainment of common goals, and the prospect of technological innovations that are likely to prove helpful in problem solving. Settings involving large numbers of actors for whom attaining common goals would necessitate major sacrifices, for example, are not conducive to the use of goal setting as a governance strategy. It is too easy for individual actors to conclude that whatever they do will not make a significant difference in such settings. Technological innovations, on the other hand, can alleviate problems that loomed large prior to the innovations (for example, dealing with congestion in uses of the geomagnetic spectrum, tracking compliance on the part of individual states or private actors). In such cases, goal attainment may prove much easier than expected at the time of the articulation of the goals. Beyond this lie issues of community, culture, and shared history in the pursuit of common goals. Where trust based on a long history of working together to address common concerns is present, for example, goal setting may become a routine method of problem solving. Where longstanding antipathies breed distrust and efforts to work together are apt to generate misunderstandings, by contrast, goal setting rarely proves effective as a means of addressing needs for governance.

A particularly relevant consideration in some settings centers on the influence of ideology or dominant social narratives. There is some tendency to associate goal setting with centrally planned or socialist systems in which the state is not only able to establish goals or targets but also to take effective steps to allocate resources or factors of production in order to attain goals. Rule making, on the other hand, is often associated with liberal systems in which the state can establish the rules of the game but is otherwise expected to minimize intrusions into the affairs of private actors and rely on regulations applicable to all (or all members of a certain class of actors) when it becomes necessary to take action to promote the public good. It is important not to exaggerate the significance of this distinction. Consider the role of goal setting in the United States in cases like the Manhattan Project during the second world war and the Apollo Project during the 1960s, or the role of environmental regulations in centrally planned systems like China. Nevertheless, it is worth thinking carefully about the prospect that the reactions of key players to measures featuring goal setting will be influenced by ideological considerations.

Mobilization of support. Those who make use of goal setting as a governance strategy in specific situations may be more or less successful in mobilizing and sustaining the support needed to attain their goals. In part, this is a matter of building coalitions of the willing, highlighting the advantages of joining coalitions of supporters and promising various sorts of rewards for those who not only do their part in fulfilling goals but also encourage others to do so. Partly, it is a matter that involves a kind of championship or leadership on the part of influential individuals who can present goals in an appealing manner and inspire those working to attain the goals with a sense of mission regarding the importance of their roles in the process. In either case, however, it is not sufficient simply to articulate goals and expect the members of society to make a concerted effort to fulfill them. Goal setting as a governance strategy requires a concentrated effort to energize supporters to pursue the common objective, sometimes over a relatively long period of time.

Enhancing the Effectiveness of Goal Setting

What options are available for those seeking to enhance the effectiveness of goal setting as a governance strategy at the international level? The essential challenge here is to influence the behavior of relevant actors by sharpening their understanding of needs for governance, strengthening the commitments they make to pursuing key goals, and providing them with compelling motives to live up to their commitments.

In some cases, such efforts may feature incentives that can be couched in terms of a utilitarian calculus of benefits and costs. Think of some of the suggestions of analysts like Schelling regarding committal tactics in situations where incentives to defect may be strong and credibility problems are therefore prominent (Schelling 1960). An actor may agree to incur costs in the event of failure to live up to a pledge, for instance, as a means of making its commitment real to itself and credible to others. There are also cases involving incentives that seem more credible to some than to others. Religious organizations, for example, regularly persuade the faithful to live up to demanding commitments (for example, tithing) by assuring them that doing so will entitle them to enjoy rewards in heaven. While this mechanism will not appeal to those who do not believe in an afterlife, it is undeniably effective in guiding the behavior of believers.

On the other hand, there are cases in which techniques for enhancing the effectiveness of goal setting rely on mechanisms that are hard to frame in terms of benefit/cost calculations. This is especially true in situations involving collective-action problems where there are the familiar incentives to defect or become a free rider (Olson 1965). Here, it may make sense to rely on mechanisms that involve factors like honor, moral obligation, face saving, a sense of group solidarity, or even the force of habit.

What measures are likely to increase the effectiveness of goal setting in the anarchic setting of international society where many of the conventional procedures for enhancing the effectiveness of goal setting are of limited value? Here are some observations about a range of procedures (by no means mutually exclusive) that may prove helpful in this setting.

Publicize the goals in dramatic forms. A goal like halving "the proportion of people whose income is less than $1 a day" by 2015 (Millennium Development Goal 1) lends itself to formulation as a sound byte that is easy to understand and to present as a challenge to one and all. It is also easy to devise a visual barometer that allows everyone to follow progress (or lack of progress) toward fulfilling such a goal on a continuous basis and to determine whether benchmarks along the way are being met. Compare this case with the goal of avoiding "dangerous anthropogenic interference with the climate system," which involves an imprecise target and no obvious measurement procedure. As recent experience makes clear, it is even possible for some actors to gain traction by denying that the problem of climate change is a consequence of human actions.

Memorialize the goals in a high-profile document or declaration. Although UN General Assembly resolutions are not legally binding, they can become high-profile documents that serve to increase the visibility of goals, to provide key goals with a sense of legitimacy, and to draw attention to the extent to which actors contribute toward attaining them. The UN General Assembly's Millennium Declaration launching the Millennium Development Goals in 2000 is a prominent example (UNGA 2000). So is the 2015 resolution formalizing the set of Sustainable Development Goals (UNGA 2015).

Formalize commitments. It may help to formalize commitments even when they are not legally binding. Fundraisers are familiar with this mechanism. Those who agree to contribute US$100 a month to some worthy cause are likely to get into the habit of doing so on a regular basis, even when they are not legally obligated to make good on these commitments. An important principle of fundraising is that those who have already started to give are the most likely candidates to provide additional contributions. They may even routinize their contributions by authorizing automatic charges to bank accounts or credit cards. In such cases, it will take a conscious act to depart from the path leading toward goal attainment; path dependence then works in favor of success.

Make formal pledges so that nonperformance will cause embarrassment or loss of face. The idea here is that actors are likely to avoid the embarrassment of reneging on pledges, even when they are under no formal obligation to make good on them. The pledges that countries were called upon to make under the terms of the 2009 Copenhagen Accord on climate change provide a prominent example of this mechanism (Copenhagen Accord 2009). Many have criticized the Copenhagen Accord on the grounds that these pledges are essentially voluntary in character. Yet it is interesting to observe the extent to which leaders feel some sense of obligation about at least attempting to fulfill their pledges.

Launch a social movement dedicated to attaining the goal. While protecting the climate system from dangerous anthropogenic interference may be an amorphous goal, the goal of the movement calling itself 350.org is understandable to all; it is easy to monitor movement either toward or away from attaining this goal. Equally important, the goal of limiting the concentration of carbon dioxide in the atmosphere to 350 parts per million has become the rallying cry for a social movement galvanizing the actions of people around the world (McKibben 2013). Whatever their merits from the perspective of rational choice, the actions of social movements can trigger profound changes in societies (for example, the abolition of slavery, the granting of universal suffrage), especially when the goals of such movements are easy to understand and progress toward goal attainment is unambiguous.

Make the goals legally binding. Giving goals the force of law may increase the willingness of actors to live up to their commitments or pledges, on the grounds that legal obligations have a normative pull of their own. The idea here is that making actions legally mandatory can influence behavior even when there are no formal sanctions or the penalties for noncompliance are modest. The goal of avoiding "dangerous anthropogenic interference with the climate system," articulated in article 2 of the UN Framework Convention on Climate Change, is a prominent example. Yet it is important to note as well that the behavioral significance of a sense of legal obligation is influenced by broader cultural perspectives that are likely to vary from place to place and across time.

Establish well-defined benchmarks for assessing progress. In addition to developing indicators for measuring progress, it often helps to establish an explicit timetable for making progress toward reaching goals and to define benchmarks to be used in assessing whether efforts to attain goals are on track. Such measures have the effect of subdividing the overall goal into more manageable chunks and establishing checkpoints that facilitate efforts to assess progress and allow for mid-course corrections if necessary. Especially in cases where achieving an overall goal is apt to be a lengthy process, some actors will find it helpful to have definite benchmarks that can serve as interim targets in addition to the final goal.

Tie other goals or the receipt of rewards to the fulfillment of the goals. Yet another procedure is to make fulfillment of initial goals a requirement for going on to the pursuit of higher-order and highly valued objectives. This is a common procedure at the individual level, where promotion or advancement to a higher status or rank is commonly predicated on the fulfillment of more or less specific goals treated as qualifiers. Think of familiar situations in which passing specific courses is treated as a prerequisite for enrollment in higher-level courses. At the international level, an interesting illustration of this mechanism is the requirement of meeting various intermediate goals as a precondition for admission to membership in the European Union.

Implications for the Sustainable Development Goals

What inferences can we draw from this general analysis of goal setting as a governance strategy that may prove helpful, not only to those concerned with conceptualizing the Sustainable Development Goals, but also to those responsible for moving the goals from paper to practice during their intended lifespan from 2016 through 2030? Just as analyses of the effectiveness of rule making emphasize the issue of fit or the need to achieve a good match between the nature of the problem at hand and the character of the institutional arrangements created to solve it, goal setting needs to be tailored to the circumstances prevailing in specific settings (Galaz et al. 2008).

At the outset, it is useful to draw a distinction between the Millennium Development Goals and the Sustainable Development Goals as goal-setting exercises. The idea of the Millennium Development Goals arose during the 1990s as a means of responding to the concerns of developing countries during a period featuring an emphasis on issues like climate change and biodiversity of particular interest to advanced industrial countries. This accounts for the emphasis in the Millennium Development Goals on concrete issues, such as eradicating poverty, improving sanitation, and providing primary education, of special concern to developing countries. In effect, the Millennium Development Goals constituted one side of a global political bargain (Young and Steffen 2009).

In an important sense, the effort to develop and implement Sustainable Development Goals is a different and more ambitious proposition. As the efforts of the Open Working Group created to follow up on the mandate to formulate Sustainable Development Goals from the 2012 UN Conference on Sustainable Development suggest, many participants have sought to emphasize continuity between the Millennium Development Goals and the Sustainable Development Goals. Critical problems involving poverty, food security, basic human health, and so forth certainly have not gone away. Nevertheless, the fundamental challenge in formulating the Sustainable Development Goals is to find a way to balance these ongoing concerns with growing systemic challenges in order to make progress toward integrating the social, economic, and environmental elements of sustainable development under conditions in which the impacts of human actions are significant at the planetary level (Young et al., this volume, chapter 3). This poses a problem in part because there is no consensus regarding the meaning of sustainable development itself at the operational level, much less regarding the implications of the onset of the Anthropocene for the pursuit of sustainable development. But what is clear already is that formulating and implementing the Sustainable Development Goals will require a global deal acceptable to both the developing and the industrialized countries (Stern 2009). The politics of the situation may stymie efforts to agree on the terms of such a deal or produce an agreement whose terms are too vague to provide useful guidance to policymakers. Nevertheless, this is not a valid reason to avoid making a concerted effort to meet this challenge.

The outcome document from the 2012 UN Conference on Sustainable Development states that the Sustainable Development Goals "should be action-oriented, concise and easy to communicate, limited in number, aspirational, global in nature and universally applicable to all countries while taking into account different national realities, capacities and levels of development and respecting national policies and priorities" (UN Conference on Sustainable Development 2012, par. 247). What does the analysis of goal setting as a governance strategy presented in this chapter have to say about fulfilling these requirements?

To lend substance to my response to this question, I will refer to the recommendations set forth in two prominent reports—the report of the High-level Panel of Eminent Persons on the Post-2015 Development Agenda, "A New Global Partnership: Eradicate Poverty and Transform Economies Through Sustainable Development" (hereafter the High-level Panel report); and the report of the Leadership Council of the Sustainable Development Solutions Network, "An Action Agenda for Sustainable Development" (hereafter the Sustainable Development Solutions Network report)—as sources of examples (High-level Panel of Eminent Persons 2013; Leadership Council for the Sustainable Development Solutions Network 2013). Of course, these are not the only prominent examples of thinking about the formulation of the Sustainable Development Goals; quite the contrary. But they are high-profile contributions to the public debate regarding the Sustainable Development Goals, and they are helpful for purposes of illustrating the concerns addressed in this section.

Minimize the number of distinct goals. Those working to develop and implement the Sustainable Development Goals do not have the luxury of the leaders of universities or charitable organizations, who can launch a fundraising campaign with the single goal of raising some specified sum of money. Not only is sustainable development a multidimensional concept; there are also numerous political pressures to include objectives of particular interest to a range of influential stakeholders. Nevertheless, my analysis of goal setting suggests that there are compelling reasons to heed the injunction of the outcome document from the 2012 UN Conference on Sustainable Development to make goals like the Sustainable Development Goals "limited in number." Articulating a suite of goals that are both numerous and range across a broad array of issue areas is bound to lead to competition for priority attention and conflict over the allocation of scarce resources. Consider in this context the High-level Panel report's 12 goals ranging from ending poverty and securing sustainable energy to ensuring stable and peaceful societies, or the Sustainable Development Solutions Network report's 10 goals ranging from achieving development within planetary boundaries to curbing human-induced climate change and securing ecosystem services and biodiversity. It is not hard to comprehend the processes that produced these formulations. But there is little prospect of making significant progress across suites of goals that encompass a large proportion of the overarching set of human interests and aspirations. It will be necessary to do better than this to make the Sustainable Development Goals effective. For comparison, the Open Working Group's draft circulated initially during the summer of 2014, and now the basis of the General Assembly's resolution on the Sustainable Development Goals, includes 17 distinct goals ranging from eradicating poverty to achieving peace and justice for all (Open Working Group 2014; UNGA 2015).

Strike a balance between aspirations and political feasibility. Goals that are lacking in ambition may be comparatively easy to fulfill, but they are not capable of galvanizing political will on the part of societies to mount the sort of campaign needed to solve fundamental problems. Conversely, goals that are too idealistic or visionary will strike actors as beyond the realm of what is politically feasible and fail to serve as the unifying themes needed to make real progress. This is why a goal like eradicating poverty is appealing. It is obviously ambitious, but it also seems to lie within the realm of the possible, especially given the progress that has been made in addressing the problem of poverty within the framework of the Millennium Development Goals. As a number of observers have argued, the period 2016–2030 may be the moment to finish the job when it comes to putting an end to extreme poverty. The High-level Panel report makes this the focus of its first proposed goal. On the other hand, fulfilling some of the goals articulated in the High-level Panel report and the Sustainable Development Solutions Network report would require a sea change in human affairs that is difficult to foresee during the 2016–2030 time frame. The idea of ensuring stable and peaceful societies is a case in point. So is the goal of devising effective measures to protect ecosystem services. There is nothing wrong with goals of this sort as long-term visionary objectives. But it is hard to see how they strike a suitable balance between aspirations and political feasibility for the period from 2016 to 2030.

Devise effective procedures to track progress. Once again, the contrast with the example of the capital campaign is instructive. Such a campaign has a single goal that is inherently operational. There is no need to devise an elaborate apparatus of targets and indicators to track progress toward fulfilling such a goal. In fact, it is possible to present a single chart using an analogy to a barometer to display progress toward reaching the overall goal on a day-to-day basis. It is also relatively easy to establish temporal benchmarks in such cases. Naturally, things are more complex when it comes to formulating and implementing the Sustainable Development Goals. Still, it is important to bear in mind the injunction from the 2012 UN Conference on Sustainable Development to make the goals "concise and easy to communicate." This is one reason why ending poverty is an appealing goal. So long as we provide an operational definition of poverty (for example, living on less than US$1 or $1.25 a day), it is comparatively easy to track progress toward fulfilling this goal. But other goals, ranging from achieving equitable growth to ensuring good governance and transforming governance for sustainable development, present fundamental challenges for those seeking to track progress. Partly, this is a matter of constructing operational measures, a situation that explains the elaborate and rather cumbersome effort to construct suites of targets and indicators to go along with each of the Sustainable Development Goals (see Pintér, Kok, and Almassy, this volume, chapter 5). In part, however, it presents fundamental challenges that are more normative in nature, such as determining what we mean by equity or good governance (see Biermann et al., this volume, chapter 4).

Make goals attractive to different motives driving behavior. Because behavior has a variety of sources, it is important to formulate the Sustainable Development Goals in a manner that appeals to those whose behavior is rooted in a range of motives. One useful distinction in this context differentiates the logic of consequences, featuring incentives linked to calculations of benefits and costs, from the logic of appropriateness, marked by more normative concerns and matters of principle (March and Olsen 1998). Especially in cases where the goals address collective-action problems (for example, protecting the earth's climate system) or the need to avoid unintended (and often unforeseen) side effects (for example, damages to ecosystem services arising from efforts to promote food security), it is important to find ways to proceed that encourage actors to transcend narrow conceptions of self-interest and to embrace principles (for example, the precautionary principle or the polluter pays principle) that can motivate all parties to behave in ways that promote the common good, even when such behavior may prove costly in the short run (Young 2001). This suggests an important rationale underlying familiar goals like providing high-quality education, ensuring healthy lives, and promoting food security (Goals 3, 4, and 5 of the High-level Panel report). There is an important sense in which all should embrace these goals on the basis of self-interest. Even the wealthy will find that meeting these goals is a matter of enlightened self-interest, since doing so will contribute over time to securing a safe, vibrant, and productive society that is beneficial to all.

Join goal setting and rule making to create integrated and effective governance systems. There is much to be said for finding ways to join goal setting and rule making to maximize the effectiveness of governance systems. Goal setting serves an aspirational function, providing actors with vision and a guiding rationale for participating in a governance system. Rule making, on the other hand, can provide the behavioral prescriptions (that is, requirements and prohibitions) needed to tell actors how they need to behave to make progress toward fulfilling goals. Goals in the absence of rules are apt to degenerate into vague aspirations that everyone embraces conceptually but no one knows how to fulfill in practice. Rules in the absence of goals, on the other hand, are apt to degenerate into burdensome and bureaucratic requirements that no one sees as needed to achieve overarching goals. This suggests a need to do more to connect goal setting and rule making as governance strategies in the global effort to pursue sustainable development. At this stage, the effort to pursue the Sustainable Development Goals is proceeding on a separate track with little input from those working on issues like the control of disease vectors, the reduction of greenhouse gas emissions, or the protection of endangered species. This is not to say that the effort to formulate and implement a set of Sustainable Development Goals for the period 2016–2030 is misguided. But to the extent that this process operates without strong links to efforts to address a variety of substantive issues, the prospects for achieving the buy-in required to make real progress toward fulfilling suitably ambitious goals will suffer.

Conclusion

The pursuit of the Sustainable Development Goals is fraught with pitfalls. But it also provides an opportunity to differentiate between goal setting and rule making as distinct governance strategies, and to examine both the conditions under which each is likely to prove effective and the prospects for combining them in a manner that produces synergy. In an important sense, the process regarding the Sustainable Development Goals differs significantly from the process in the 1990s that led to the Millennium Development Goals. Whereas the earlier process reflected a political bargain designed to engage developing countries and persuade them to join efforts to tackle issues of global environmental change (for example, climate change or the loss of biological diversity), the later process is about devising an approach to the full range of human-environment interactions on a human-dominated planet and finding ways to track accomplishments in this realm (Young and Steffen 2009). There is no guarantee that this process will yield useful results; it may well have produced a list of goals that is too long and framed in terms that are too vague to provide practical guidance. But the process does offer an opportunity to chart a global course for living sustainably during an era in which 7 to 9 billion human beings have achieved the capacity to dominate planetary systems.

References

Chayes, Abram, and Antonia Handler Chayes. 1995. *The New Sovereignty: Compliance with International Regulatory Agreements*. Cambridge, MA: Harvard University Press.

Galaz, Victor, Per Olsson, Thomas Hahn, Carl Folke, and Uno Svedin. 2008. The Problem of Fit among Biophysical Systems, Environmental and Resource Regimes, and Broader Governance Systems: Insights and Emerging Issues. In *Institutions and Environmental Change*, ed. Oran R. Young, Leslie A. King and Heike Schroeder, 147–186. Cambridge, MA: MIT Press.

High-level Panel of Eminent Persons on the Post-2015 Development Agenda. 2013. *A New Global Partnership: Eradicate Poverty and Transform Economies Through Sustainable Development.* New York: United Nations.

Karabell, Zachary. 2014. *The Leading Indicators: A Short History of the Numbers that Rule the World.* New York: Simon and Schuster.

Leadership Council for the Sustainable Development Solutions Network. 2013. An Action Agenda for Sustainable Development. Report to the UN Secretary-General.

March, James G., and Johan P. Olsen. 1998. The Institutional Dynamics of International Political Orders. *International Organization* 52: 943–969.

McKibben, Bill. 2013. *Oil and Honey: The Education of an Unlikely Activist.* Collingwood, Austria: Black Inc.

Olson, Mancur Jr. 1965. *The Logic of Collective Action.* Cambridge, MA: Harvard University Press.

Open Working Group on Sustainable Development Goals. 2014. Outcome document. Available at: http://sustainabledevelopment.un.org/focussdgs.html.

Sachs, Jeffrey D. 2015. *The Age of Sustainable Development.* New York: Columbia University Press.

Schelling, Thomas C. 1960. *The Strategy of Conflict.* Cambridge, MA: Harvard University Press.

Stern, Nicholas. 2009. *The Global Deal: Climate Change and the Creation of a New Era of Progress and Prosperity.* New York: Public Affairs.

UN, United Nations. 1992. Framework Convention on Climate Change. Available at: http://unfccc.int/.

UN, United Nations. 2012. The Future We Want. Outcome document from the UN Conference on Sustainable Development. Res. 66/288.

UNGA, United Nations General Assembly. 2000. Millennium Declaration. UN Res. 55/2. Available at: http://www.un.org/millennium/declaration/ares552e.htm.

UNGA, United Nations General Assembly. 2015. Transforming Our World: The 2030 Agenda for Sustainable Development. Draft resolution referred to the United Nations summit for the adoption of the post-2015 development agenda by the General Assembly at its sixty-ninth session. UN Doc. A/70/L.1.

United Nations. 2009. Framework Convention on Climate Change. Copenhagen Accord. Outcome document from fifteenth Conference of the Parties. Available at: http://unfccc.int.

Young, Oran R. 2001. Environmental Ethics in International Society. In *Principles of Ecosystem Stewardship*, ed. Jean-Marc Coicaud and Daniel Warner, 161–193. Tokyo: UNU Press.

Young, Oran R. 2011. The Effectiveness of International Environmental Regimes: Existing Knowledge, Cutting-edge Themes, and Research Strategies. *Proceedings of the National Academy of Sciences of the United States of America* 108: 19853–19860.

Young, Oran R., Leslie A. King, and Heike Schroeder, eds. 2008. *Institutions and Environmental Change*. Cambridge, MA: MIT Press.

Young, Oran R., and Will Steffen. 2009. The Earth System. In *Principles of Ecosystem Stewardship*, eds. F. Stuart Chapin, Gary Kofinas, and Carl Folke, 319–337. New York: Springer.

3 Goal Setting in the Anthropocene: The Ultimate Challenge of Planetary Stewardship

Oran R. Young, Arild Underdal, Norichika Kanie, and Rakhyun E. Kim

The UN General Assembly's Millennium Declaration, adopted in 2000, launched a global effort to eradicate poverty, improve basic human health, and enhance food security, educational opportunities, and gender equality. Although exogenous factors, such as economic growth and democratic reforms, have played important roles in the progress made since the adoption of the Millennium Development Goals, experience with the pursuit of the Millennium Development Goals has stimulated interest in goal setting, in contrast to rule making, as a strategy for solving global problems (Haas and Stevens, chapter 6, Andresen and Iguchi, chapter 7, and Yamada, chapter 8, all this volume). The call for Sustainable Development Goals in "The Future We Want," the outcome document from the 2012 UN Conference on Sustainable Development, is a clear expression of this growing interest in governance through goals (UNGA 2012).

As the effort to craft the terms of a broadly agreeable set of Sustainable Development Goals has made clear, interest in the concerns underlying the Millennium Development Goals remains strong. Throughout the process, issues relating to poverty, hunger, health, education, and gender equality headed the lists produced by both official and unofficial contributors. Yet framing and specifying a set of Sustainable Development Goals is not simply a matter of rededicating the global community to addressing these familiar concerns. Sustainable development is a broader objective that calls for a melding of economic, social, and environmental factors, both to enhance the well-being of individual humans and to produce resilient socio-ecological systems from the local to the global level. Today, scientists as well as a growing number of policy makers are increasingly aware that the earth itself has become, over a short span of time, a human-dominated system (Steffen et al. 2004). The resultant growth of a new discourse, often framed in terms of the proposition that the earth is entering a new era referred to as the Anthropocene, has profound consequences for our thinking about the pursuit of sustainable development at all levels.

In this chapter, we discuss the transition from the Millennium Development Goals to the Sustainable Development Goals, identify major features of the Anthropocene that have important implications for goal setting as a governance strategy, illustrate the implications of this new way of thinking through a case study dealing with freshwater, and explore reasons why the effort to fulfill the Sustainable Development Goals must meet challenges that are greater than those confronting the Millennium Development Goals. We conclude by considering the proposition that developing a sustainability *Grundnorm* may prove helpful to those responsible for implementing the Sustainable Development Goals between 2016 and 2030.

Moving from the Millennium Development Goals to the Sustainable Development Goals

There is considerable overlap in the content of the Millennium Development Goals and the Sustainable Development Goals. Both sets of goals stress the importance of alleviating extreme poverty, eradicating major diseases, and taking steps to promote gender equality. Whereas the Millennium Development Goals marked the beginning of a new era by laying out an ambitious agenda focusing on issues of particular importance to developing countries, however, the Sustainable Development Goals highlight challenges that require substantial behavioral changes on the part of residents of developed countries as well as efforts to improve the circumstances of those living in developing countries. The critical shift is embedded in the idea of sustainable development. In adding the modifier "sustainable" to the goal of "development," the 2012 UN Conference on Sustainable Development directed attention to the proposition that future progress in meeting human needs and aspirations requires a strong commitment to safeguarding the earth's life-support systems.

In adding "sustainable," the United Nations also called attention to previous achievements. Twenty-five years earlier, the World Commission on Environment and Development (popularly known as the "Brundtland Commission") told us, in language that has become iconic, that *sustainable* development is "development that meets the needs of the present without compromising the ability of future generations to meet their own needs" (World Commission on Environment and Development 1987). Measuring progress toward sustainable development conceptualized in this way has proven to be a difficult task. Many have proposed methods for operationalizing the concept of sustainable development, but no consensus has emerged on the merits of a particular set of procedures for operationalizing this goal. Nevertheless, framing the issue as a matter of sustainable development not only stresses the importance of meeting human needs on a lasting basis; it also draws attention to the fact that human needs extend far beyond what is captured in conventional measures of income or wealth (Sachs 2015).

In summarizing conclusions from two decades of research under the International Geosphere-Biosphere Programme, Steffen and colleagues observed that human activities have become "equal to some of the great forces of nature in their extent and impact" (Steffen et al. 2004, 257). As a consequence, they added, the earth system is now operating in "a no-analogue state," meaning that previous experience may no longer be a reliable guide to the future (Steffen et al. 2004, 262). This transformation, which Steffen and colleagues labeled the "great acceleration," presents us with a series of unprecedented challenges that any effort to achieve sustainable development on a planetary scale must confront. It also makes clear that the pursuit of sustainable development must be guided by improved means of managing or steering individual and collective human behavior that now heavily influences the fate of the planet, and with it the course of both individual and social welfare. Some recently proposed definitions of sustainable development explicitly include the protection of the earth's life-support systems as a defining characteristic (Griggs et al. 2013; Muys 2013; Sachs 2015).

The interaction of biophysical and socioeconomic changes has brought about the closing of the planetary frontier. We can no longer move on to new stocks of fish, new forests, or new reserves of fossil fuels when old ones give out; we must learn to live within our collective means on a finite planet with a rapidly rising human population (Berkes et al. 2006). In this setting, human actions occurring in one place can have significant effects in far-away places. Greenhouse gases emitted mainly in the densely populated societies of the mid-latitudes, for example, are major drivers of the dramatic melting of sea ice in the Arctic, the sharp rise in the acidity of the oceans threatening the survival of coral reefs in the tropics, and the loss of species in the rainforests of the Amazon basin. Increasingly, biophysical systems are reaching thresholds or tipping points where small trigger events can cause far-reaching changes that are nonlinear, irreversible, and sometimes abrupt in nature (Lenton et al. 2008). As a result, extreme events that we are unable to predict and that consequently take us by surprise are occurring more frequently. The 2011 tsunami triggered by the Great East Japan Earthquake, which then caused the disaster at the Fukushima Daiichi nuclear power plant, is a particularly dramatic example of an increasingly important class of events featuring such chain reactions with impact regardless of national borders. Success in addressing the Sustainable Development Goals, then, must encompass a strategy for coming to terms with these large-scale, even planetary, processes, while at the same time continuing to make progress in solving longstanding problems like alleviating extreme poverty and eradicating debilitating diseases.

Why Does the Onset of the Anthropocene Matter?

Beginning roughly at the midpoint of the twentieth century, a series of socioeconomic and biophysical developments converged to launch the "great acceleration" (Steffen et al. 2004; Young 2013). As a consequence of this convergence, the earth as a whole has become a human-dominated system. Unfolding over a period of several decades, this transition has entered our thinking about sustainable development through various channels, including the series of assessment reports from the Intergovernmental Panel on Climate Change and via increasing media attention to the introduction and rapidly growing use of the concept of the Anthropocene. Specific features of the Anthropocene that have far-reaching implications for efforts to meet needs for governance in general and for the use of goal setting as a governance strategy in particular include the rise of teleconnections, the emergence of planetary boundaries, an increased incidence of nonlinear changes, and the growing importance of emergent properties on a global scale (Cornell, Prentice, House, and Downy 2012).

Teleconnections. Teleconnections are systemic linkages that connect widely separated and seemingly unrelated events in the earth system. Many teleconnections are biophysical. Rising emissions of greenhouse gases in the mid-latitudes are causing the collapse of sea ice in the Arctic and the bleaching of corals in the tropics. Others are socioeconomic. Financial problems arising in specific settings (for example, the bursting of the housing bubble in the United States in 2008) can trigger chain reactions that have global consequences. As a result, efforts to pursue goals in one setting can be disrupted or enhanced by occurrences in distant locations or in seemingly unrelated sectors. The implication of this development for goal setting is twofold. First, goals must be "global in nature and universally applicable to all countries while taking into account different national realities, capacities and levels of development" (UNGA 2012, par. 247). Of course, eradicating poverty is a goal that has spatial coordinates; there are more poor people located in some parts of the world than in others. Increasingly, however, critical goals like ensuring food security and dealing with climate change require programmatic efforts that are global in scope. Second, it is essential to consider linkages between goals that may appear at first glance to be unrelated. The Open Working Group of the General Assembly on Sustainable Development Goals sought to address this matter, even though its efforts were not always successful. To be successful in the long run, the Goals will need to take advantage of synergies in addition to avoiding conflicts.

Planetary boundaries. With a human population of over 7 billion expected to grow to 9 billion by 2050, the limits of the earth as a home for human beings are becoming increasingly clear (Rockström et al. 2009; Steffen et al. 2015). Anthropogenic drivers are causing the depletion of the planet's natural resources and the disruption of the planet's great cycles (for example, the carbon and nitrogen cycles) at a rate that is unprecedented. The implication of this development is that margins for error are decreasing. We must replace the practices of "roving bandits," who deplete specific resources (for example, fish stocks) and then move on to repeat the process in new territories, with practices that allow for the continued use of atmospheric, marine, and terrestrial systems in a manner that is sustainable over the long term (Berkes et al. 2006). What this suggests is that we need to think in systemic terms and to recognize that a critical goal will be the adoption of a guiding discourse of stewardship, whether we are dealing with food, water, energy, or any other specific human need (Chapin, Kofinas, and Folke 2009).

Nonlinearities. Increasingly, we are faced with changes that are dramatic, abrupt, and irreversible rather than incremental, gradual, and reversible. An important feature of this aspect of the Anthropocene is the emergence of tipping points or thresholds where relatively modest trigger events can touch off cascades of change that ripple through large systems (for example, the earth's climate system) with consequences that are far-reaching and irreversible on a human time scale (Lenton et al. 2008). An important goal, under the circumstances, is to make a concerted effort to identify tipping points sufficiently far in advance to steer human activities away from crossing thresholds and to put in place self-correcting or negative feedback mechanisms that kick in when the system approaches a point of no return.

Whereas reacting after the fact was once sufficient, we must now react before the fact, and on a planetary scale. The Sustainable Development Goals must target behavioral change among members of the current generation rather than focusing on the actions of future generations.

Emergent properties. One critical insight arising from the preceding observations regarding the Anthropocene is the realization that the earth and its major systems (for example, the climate system) are highly complex. We are regularly taken by surprise by the consequences of seemingly modest actions that trigger large-scale and unforeseen results. The climate system is a prominent example. The earth system may well encompass negative feedback mechanisms that will serve to moderate the effects of rising concentrations of greenhouse gases in the atmosphere. At least as likely, however, is the prospect that positive feedback mechanisms will act to broaden and intensify the impacts of anthropogenic drivers. What this means is that there is no escape from decision-making under uncertainty (Kahneman 2011). We must avoid paralysis in the face of our limited understanding of complex systems. But this development does suggest that there is a need to set goals in a manner that leaves room for rapid adjustments in the face of systemic surprises.

An Example: The Quantity and Quality of Freshwater

What do these planetary developments mean in practice for those seeking to make use of goal setting as a global governance strategy? To translate the analytic observations of the preceding section into practical considerations, we explore their implications for the case of freshwater. It would be easy to make similar observations regarding other major issues like feeding 10 billion people, weaning industrial societies off fossil fuels, or reducing the pressures driving more and more species to extinction. But the case of freshwater is attractive for several reasons. Many knowledgeable observers expect challenges relating to freshwater to become *the* defining issue of the twenty-first century on a global scale. While securing adequate supplies of freshwater is an age-old problem on a local scale, issues pertaining to freshwater have emerged as priority concerns on a regional level and even at the global level. There is as well a powerful nexus linking freshwater to a variety of other major concerns, including the achievement of food security, the fulfillment of urgent needs for improved sanitation, and the generation of adequate supplies of energy. Later chapters in this volume pick up on the case of freshwater in a variety of substantive applications (see Haas and

Stevens, chapter 6; Yamada, chapter 8; and Gupta and Nilsson, chapter 12, all this volume).

There is enough accessible freshwater on the planet to meet the needs of the present world population (Gleick et al. 2014). The critical challenges regarding water pertain to (a) the distribution of water or the distances separating areas of dense human population and the location of major freshwater reservoirs, (b) the economic and political mechanisms we have developed over time to govern the allocation of scarce water resources among human users, and (c) the unintended impacts of a wide range of human activities on water quality. As is the case with other priority concerns like food security, therefore, a Sustainable Development Goal focusing on water must direct attention to matters of distribution, allocation, and contamination rather than dealing only with the matter of quantity.

The distribution of water on the planet is highly uneven. Canada has a relatively small human population, but an abundance of water. While South China has an adequate supply of water, the northern and especially the northwestern parts of the country suffer from severe water shortages and increasing desertification. In the United States, the West and especially the Southwest are arid regions where water shortages have reached crisis proportions as a result of several years of severe drought. Moreover, the consequences of climate change are expected to exacerbate these disparities. Worldwide, the general expectation is that the impact of climate change will make wet areas wetter and dry areas dryer. In specific areas, regional developments may intensify the impacts of these developments. In Asia, for example, the disruption of the Asian monsoon system is a distinct possibility, and saltwater intrusions resulting from rising sea levels may drown large coastal areas and contaminate freshwater supplies in countries like Bangladesh. In Europe, extreme flooding of the major rivers has become routine, whereas water flowing down the Colorado River in western North America has reached record lows in recent years.

Some strategies designed to achieve water security under these conditions involve massive engineering projects and high politics. The efforts of Los Angeles, a rapidly growing city situated in an arid environment, to enhance its water supply by buying water rights in distant locations and transporting the water over long distances through aqueducts have become legendary (Reisner 1993). China is expending the equivalent of tens of billions of dollars on one of the most massive engineering projects in human history, designed to move large quantities of water from the more-endowed south to the arid north (Kuo 2014). To address the problem of the availability of water, planners in arid countries like Israel have made major commitments to desalinization, despite the fact that meeting needs for freshwater in this fashion is both expensive and energy-intensive. An alternative strategy for addressing this problem involves promoting the use of "green" water, a system in which food and other water-intensive commodities are produced in areas where water is plentiful and marketed in arid regions in return for products that are better suited to production in areas in which water is in short supply. But this strategy requires dependable conditions of peace that allow commerce to thrive. It is technically feasible to reclaim water on a large scale, treating it in ways that make it reusable for many purposes, including human consumption. However, a widespread antipathy to the use of reclaimed water, based largely on fear or prejudice in contrast to rational calculations, constitutes an obstacle to the use of this strategy on a large-scale basis.

Antiquated or underdeveloped institutional arrangements make it difficult to deal with many of these challenges. In California, for example, some 85% of the surface water and most of the groundwater goes to agriculture, which accounts for only 3% of the state's economy, despite the availability of technologies (for example, various forms of drip irrigation and techniques of dry farming) that would make it possible to reduce water consumption dramatically without any loss of production (Bardach 2014). The problem lies largely in the fact that the US federal government subsidizes agricultural users of water, who therefore lack incentives to invest in the appropriate technologies. Nor is this situation limited to the United States. Worldwide, some 70% to 90% of available freshwater is used to irrigate land for agriculture (Scanlon et al. 2007). But inefficient agriculture is by no means the only source of excessive uses of water. There are great opportunities in urban and suburban settings to reduce the demand for water by introducing more efficient systems of water use or adjusting lifestyles that are insensitive to the threats of water shortages. For the most part, however, incentives to conserve are weak and users are resistant to calls for adjusting their water-intensive lifestyles. Sometimes, needed management regimes are simply absent. The Ogallala Aquifer, for example, contains a massive reservoir of groundwater underlying a large swath of the mid-section of the United States. But because it is a common pool resource whose appropriators are not subject to a unified, much less effective, regulatory system, users are exploiting this resource at a pace far exceeding the natural recharge rate. In effect, they are mining fossil water on an unsustainable basis (Little 2009).

In other cases, policies and practices regarding uses of water are tightly connected to issues of national security. Both the flooding of northern Sudan resulting from the construction of the Aswan Dam in Egypt in the 1950s and the more recent controversies relating to reductions in water from the Euphrates River reaching Iraq as a consequence of the construction of the Atatürk Dam in Turkey are prominent examples. A common argument against increased reliance on "green" water comes from those who believe arrangements of this type will compromise the national security of countries that become dependent on others as sources of essential commodities. Similar problems arise in cases like the Arab-Israeli conflict, where competition for the limited supplies of water in the Sea of Galilee (Lake Tiberias) and the Jordan River has long been an intractable element of the conflict. In such cases, the intrusion of high politics rules out solutions to water problems that are perfectly feasible from a technical perspective.

Even in situations where adequate quantities of water are available, issues of water quality are often prominent. For the most part, this is a matter of dealing with the unintended byproducts or side effects of activities undertaken for legitimate purposes. A classic example involves runoffs of nitrogen and phosphorus arising from inefficient or unregulated uses of chemical fertilizers in support of agricultural production. Large bodies of freshwater (for example, Lake Tai in China and Lake Champlain in the United States) now suffer from algal blooms that periodically make their waters unfit for human consumption and that lead to recurrent closures of segments of these water bodies for various human uses. Similar observations are in order with respect to industrial wastes that are allowed to run into rivers without adequate requirements regarding treatment (Fagin 2013). Notorious situations in which rivers become so polluted that they literally catch on fire are not uncommon. The problem in all these cases is fundamentally political. Even where there is room to improve social welfare by eliminating or controlling these side effects, the facts that doing so is likely to produce losers as well as winners, and that the losers often have considerable capacity to resist change, make it difficult or even impossible to address these issues of water quality effectively.

The focus of the Millennium Development Goals on the issue of supplying freshwater is a matter of continuing concern. The tension between the compelling ethical argument for treating an adequate supply of freshwater as a human right and the consequences of privatization, which has driven up the price of freshwater especially in many rural areas, constitutes an unresolved challenge (von Weizsäcker, Young, and Finger 2005). We must not lose sight of this issue. Yet dealing with the issue of water in the framing of the Sustainable Development Goals is a much larger concern. Partly, this is a matter of anticipating large-scale processes like changes in the hydrology of major river systems that threaten the livelihoods of tens or even hundreds of millions of people. In part, it raises questions about institutional arrangements (or the lack thereof) that may have been adequate in the nineteenth or even the twentieth century but that lead to unacceptable outcomes in the world of today. The challenge of dealing with the quantity and quality of water is not confined to finding ways to meet the needs of the bottom billion. It is a challenge arising on all levels from the local to the global that will require fundamental changes both in antiquated institutions and in deeply engrained patterns of behavior that are just as prevalent in the first world as they are in the third world. Importantly, these challenges involve different problems in different places, and thus require solutions tailored to different circumstances, even though they need to be considered at the same time in crafting global solutions to solve global problems.

Implications for the Sustainable Development Goals

What are the implications of this analysis for the development and implementation of the Sustainable Development Goals? The outcome document from the 2012 UN Conference on Sustainable Development states that the goals "should be action-oriented, concise and easy to communicate, limited in number, aspirational, global in nature and universally applicable to all countries while taking into account different national realities, capacities and levels of development and respecting national policies and priorities" (UNGA 2012, par. 247). This is a sensible but demanding list of requirements. How can we ensure that the Sustainable Development Goals become a means to change the course of human behavior with these requirements in mind?

We can address this question by differentiating two stages of the Sustainable Development Goals: goal setting and goal achievement. In the case of goal setting, the main challenge is to frame goals in such a way that they take into account the fundamental changes in the earth system associated with the onset of the Anthropocene, while at the same time emphasizing the continued importance of concerted efforts to eradicate poverty, ensure food and water security, improve human health, and in other ways enhance the quality of life of those in developing countries, as intended by the Millennium Development Goals. With regard to goal achievement, the critical challenge is to steer or guide a wide range of human activities effectively in the directions required by the new Sustainable Development Goals. This steering challenge is extremely demanding, and more so for the Sustainable Development Goals than it was for the Millennium Development Goals. We therefore devote this section to a discussion of achieving the set of Sustainable Development Goals formulated through the UN process of agenda setting for the period 2016–2030.

Achieving sustainable development is a quintessential *transgenerational* program. As such, it has three characteristics that make it a particularly demanding governance challenge (Underdal 2010). First, time lags between mitigation measures (usually involving short-term costs) and mitigation effects (presumably bringing benefits) are long, often extending well beyond one human generation. Second, despite substantial progress in social as well as natural science over the past four to five decades, our understanding of the role of human activities in driving earth system dynamics remains incomplete and clouded by profound uncertainties. Third, some Sustainable Development Goals deal with the provision of global collective goods of a nature that links these goods to a wide range of human activities and at the same time leaves them beyond the scope of any "single best effort" solution. (We borrow this term from Barrett [2007], who refers to collective-action problems where one single actor is *capable* of providing the collective good in question through unilateral efforts, and also has a strong *interest* in having the good provided.) For reasons of space, we focus in this section on the implications of time lags. Specifically, we consider briefly three important implications: asymmetric uncertainty, time inconsistency, and extreme intergenerational asymmetry with regard to participation and political power.

Other things being equal, *uncertainty* increases the farther into the future we look. Inherent in challenges such as climate change and biodiversity loss is a profound asymmetry between our ability to calculate the costs of mitigation measures and our ability to foresee the eventual benefits of such measures. A cursory reading of government programs and public debate indicates that short-term consequences of mitigation efforts tend to be framed largely in terms of immediate costs, while long-term damage avoided is conceived of in terms of eventual benefits. Framed in these terms, uncertainty will weigh more heavily in the assessment of mitigation benefits than in the assessment of mitigation costs. Moreover, most people are likely to think of uncertainty in this context as referring only to the "negative" error margin rather than to the full range of plausible outcomes, including some that are more positive as well as some that are less positive than the average or median estimate. The greater these asymmetries, the more they will skew the cost/benefit ratio by reducing the expected net benefit of mitigation. To make matters worse, this problem may be amplified by the well-documented human tendency to react more strongly to the prospect of a given loss than to the prospect of a gain of equal size (Kahneman and Tversky 1979). To the extent that policy makers conform to this pattern, ambitious mitigation measures will face even higher hurdles than those predicted by conventional rational choice theory.

Time lags tend to affect incentive structures in other ways as well. One likely effect is known as the *time-inconsistency* problem. Time inconsistency occurs when an actor's best plan today for some future period of time will no longer be optimal when that time arrives (Kydland and Prescott 1977). An everyday example will illustrate how this mechanism works. Assume you have embarked upon a program of regular physical exercise to improve your health. Even if you firmly believe that the program as a whole will yield substantial net benefits, you need not come to a positive conclusion regarding each training session in the program. A single defection would hardly reduce long-term health benefits, but may well boost short-term well-being (for example, by avoiding exposure to heavy rain or chilling winds or by freeing up time to attend a concert). In technical terms, the cost/benefit calculus for this particular training session shows a negative balance, despite expectations of substantial gains from the overall program. The core of the time-inconsistency problem is incongruity between the cost/benefit considerations for the plan as a whole and those pertaining to individual microdecisions required to implement the plan. Such incongruity may occur even when a single individual makes *all* the decisions. It is all the more likely to occur for programs that must survive multiple shifts in governments. And the more instances of incongruity there are, the greater the risk of multiple defections from the master plan.

Another potential distortion arises from *discounting*—a procedure used to estimate the present value of future benefits and costs. Discounting normally attributes higher value to current benefits than to those occurring in the future, for two main reasons: "human impatience" and assumptions of rising incomes (Fisher 1930). Stern (2007) and Schelling (1995), among others, have challenged the relevance of conventional discounting theory to transgenerational challenges such as global climate change.

Where the consequences of current policies materialize several decades—in some cases even centuries—later, moreover, future stakeholders will not have an opportunity to voice their concerns when preferences are aggregated to choose policies. In developing the Sustainable Development Goals, we face *extreme intergenerational asymmetries* in participation and political power. As we noted above, long time lags between the adoption of measures and the realization of their benefits mean that those who are in a position to undertake effective mitigation measures must pay most of the costs but will reap only a fraction of the benefits derived from the damage averted. The smaller that fraction, the more the material self-interest of the "upstream" generation(s) will diverge from the interests of "downstream" generations. We expect this gap to be narrowed to some extent by people sincerely caring about the well-being of their children, grandchildren, and great-grandchildren. However, if Schelling is right (Schelling 1995), those who stand to benefit the most from mitigation measures will be much more "distant" not only in terms of kinship but also with regard to geographical location, ethnicity, culture, and other aspects of collective identities. Other things being equal, the willingness to pay for benefits to be reaped mostly by others will decline the greater the perceived distance to the beneficiaries. This combination of diverging incentives and extreme power asymmetry generates a severe risk of vertical disintegration of sustainable development programs or, in other words, a wide gap between a set of ambitious goals and targets articulated in the Sustainable Development Goals on the one hand and the policies and practices actually implemented to achieve these goals on the other.

Building Institutions on a Normative Foundation

What can we do to avoid or minimize this kind of disintegration? One strategy features the introduction of *institutional* arrangements, such as the appointment of ombudspersons (commissioners), guardians (trustees), or similar types of agents mandated to speak on behalf of future generations (see, for example, the Science and Environmental Health Network and the International Human Rights Clinic at Harvard Law School 2008). The office of ombudsperson or commissioner is a construct used by many countries to protect *individual* (human) rights. Fewer countries—Hungary and Israel are the pioneers—have established offices of commissioners to protect the *collective* rights of future generations. But in neither of these cases have the offices developed into long-lasting institutions. In Hungary, the tasks assigned to this office were taken over by the Office of the Commissioner for Fundamental Rights in 2012. In the case of Israel, the office of Commissioner for Future Generations was disbanded after five years. It would be premature to judge the potential role of ombudspersons or commissioners on the basis of experience with short-lived examples from a few countries. Yet it is worth noticing that, so far, institutions of this kind have a stronger track record in protecting the rights of individuals and small groups than in protecting the collective rights of future generations to enjoy the benefits of the earth's life-support systems.

Part of the explanation may lie in the fact that the two tasks differ in important respects. In conventional human rights domains, ombudspersons or trustees are dealing with fairly specific and well-defined (individual) rights written into international conventions and protocols as well as national laws and regulations. As a consequence, rights violations are often subject to detection as they occur and may be subject to ordinary legal procedures. Moreover, the victims of such violations will normally belong to living generations and are often easy to identify. By contrast, the collective rights of future generations to benefit from the earth's life-support systems normally are defined in more general terms, are harder to monitor empirically, and are more difficult to enforce using ordinary legal procedures.

At times, both national and international courts have interpreted the law as requiring intergenerational equity and granted legal standing to those seeking to represent future generations. In its advisory opinion, "Legality of the Threat or Use of Nuclear Weapons," for example, the International Court of Justice treated the impact of nuclear weapons on the well-being of future generations as an important factor (International Court of Justice 1996). But it did not call for the outlawing of nuclear weapons for this reason. Nonetheless, the Court recognized that "the use of nuclear weapons could be a serious danger to future generations" (International Court of Justice 1996, 244). It is noteworthy as well that the Court embraced a broad definition of the environment as representing "the living space, the quality of life and the very health of human beings, including generations unborn" (International Court of Justice 1996, 241). Judge Weeramantry, in his dissenting opinion, argued that "the rights of future generations ... have woven themselves into international law through major treaties, through juristic opinion and through general principles of law recognized by civilized nations" (International Court of Justice 1996, 455).

Some domestic courts have established procedural protections for future generations by granting them legal standing. In 1994, the Supreme Court of the Philippines granted standing to 44 minors to sue the government on behalf of themselves and members of future generations regarding the impacts of unsustainable logging in the country (Philippines Supreme

Court 1994). In 1999, the Supreme Court of Montana in the United States found that the environmental provisions of the state's constitution provide standing to citizens and environmental groups to sue for environmental harms to public resources (Supreme Court of Montana 1999). However, not all such attempts have been successful. In 2001, the Seoul Administrative Court of the Republic of Korea, for example, denied legal standing to a group of children who collectively filed a lawsuit to stop a government-led large-scale coastal reclamation project (Seoul Administrative Court 2001). Without doubt, the ruling of the Seoul Administrative Court reflects the norm rather than the exception among domestic courts around the world.

These observations do not warrant the conclusion that institutional reform cannot succeed in protecting the interests of future generations. Taken together, however, they do suggest that a firm normative foundation will be needed for institutional arrangements to become effective mechanisms for pursuing this goal. This suggests a second, by no means mutually exclusive, strategy built on the introduction of a basic sustainability principle or, in other words, a sustainability *Grundnorm*. Such a *Grundnorm* would guide sustainable development policies and practices by serving as a fundamental principle of law, equal to other fundamental principles such as justice, equality, and freedom (Bosselmann 2008). At present, both national legal systems and international law lack such a fundamental principle prohibiting serious or irreversible harm to the integrity of ecosystems based on the rights of future generations.

The concept of a *Grundnorm* is commonly understood as a basic norm against which all other legal norms can be interpreted and validated (Kelsen 1967). A *Grundnorm* is a foundation upon which a legal system is based. A constitution is a good example of a *Grundnorm*: It informs and justifies all elements of the legal system (Fisher 2013). Conceptually, a *Grundnorm* exists independently of a legal system, but underpins legal reasoning in the form of an inference rule. The legitimacy of a constitution, for example, does not derive from inside but outside the legal system. A *Grundnorm* is, therefore, "a matter of political ideology rather than legal ideology" (Fisher 2013, 7). This understanding differs from that of Hans Kelsen and is closer to Immanuel Kant's argument that any positive law must be grounded in a natural law of general acceptance and reasonableness to prevent pure arbitrariness (Kim and Bosselmann 2015).

Such a conceptualization makes it possible to think of the principle of sustainability as a *Grundnorm*. The existence of a sustainability *Grundnorm* rests on the proposition that respect for planetary boundaries defining the "safe operating space for humanity with respect to the Earth system" (Rockström et al. 2009, 472) constitutes a moral imperative in the Kantian sense (Kim and Bosselmann 2015). In the specific context of global governance, we may interpret the principle of sustainability as a superior norm (or goal) that gives all international regimes and organizations a shared purpose to which their more specific activities must contribute, thereby lending coherence to what otherwise might become a disparate or even internally inconsistent collection of arrangements (Kim and Bosselmann 2013). Such a usage would parallel basic norms, like the protection of human rights or the promotion of free trade, that serve as measures of the legality of state behavior in other issue domains. In the absence of a sustainability *Grundnorm* of similar weight, the idea of sustainable development lacks force as a basis for protecting future generations and the environment.

Ideally, a sustainability *Grundnorm* would encapsulate a clearly defined and globally accepted vision for long-term sustainable development *beyond* 2030. As in the Millennium Declaration, the report of the Open Working Group on Sustainable Development Goals emphasizes poverty eradication as "the greatest global challenge facing the world today" (Open Working Group on Sustainable Development Goals 2014). The need to "end poverty in all its forms everywhere" certainly deserves priority attention. But from the perspective of achieving long-term sustainable development, the protection of "planetary must-haves" (for example, climate stability and the maintenance of ecosystem services) needs to be recognized as a necessary precondition for development of any kind (Griggs et al. 2013). All ethical standpoints, including the prevailing anthropocentrism, support this proposition because the welfare of both current and future generations depends on maintaining the earth's life-support systems. Of course, this argument is not new. Similar language appears in a range of influential texts, including the 1972 Stockholm Declaration on the Human Environment, the 1982 World Charter for Nature, the 1992 Rio Declaration on Environment and Development, and "The Future We Want." But these declarations are not sufficient. They need to be backed up by the practices of states and other actors.

The establishment of a sustainability *Grundnorm* requires acceptance of the proposition that it is a core duty of states and nonstate actors alike to "conserve, protect and restore the health and integrity of the Earth's ecosystem" (Rio Principle 7). What the integrity of the earth's ecosystem will mean in the Anthropocene remains a subject of debate. But for the purpose of implementing goal-oriented governance mechanisms for sustainable development, such as the Sustainable Development Goals, it is sufficient to agree on a practical, anthropocentric definition of global ecological integrity such as the combination of the biodiversity and ecosystem processes that characterized the biosphere as a whole during the Holocene (Bridgewater, Kim, and Bosselmann 2016). Because it is the only state of the earth system that we know for sure can support contemporary society, the Holocene provides an appropriate precautionary reference point for this purpose (Steffen et al. 2011).

Adopting and implementing a sustainability *Grundnorm* would require a major reform of existing and emerging international governance systems. At the global level, the international community is in need of a new, constitution-type agreement that will redefine the relationship between humans and the rest of the community of life (Kim and Bosselmann 2015). Potential candidates for such an agreement include the Draft International Covenant on Environment and Development, which is a product of decades of work of leading scholars and practitioners with an aim to codify existing environmental law (International Union for Conservation of Nature Environmental Law Programme 2015). The draft covenant reflects the sustainability *Grundnorm* in its core in the form of a fundamental principle of respect for "nature as a whole and all life forms" as well as the "integrity of the Earth's ecological systems" (art. 2). Ultimately, a reform of the charter of the United Nations may be required (Kanie et al. 2012).

Desirable as this integrating effort is, it will take time to put this *Grundnorm* in place. But when fully embraced, such a *Grundnorm* will prove useful in several ways. It can serve as a guide to intertemporal priority setting, calling on the present generation to pay attention to the welfare of future generations, while allowing a degree of flexibility as to how exactly to balance the needs of future generations against those of the present generation. This means accepting that, on a human-dominated planet, protection of the earth's life-support systems will be an essential component of any effective strategy for providing future generations with opportunities equivalent to those the current generation enjoys. A sustainability *Grundnorm* can also guide the interpretation of existing laws and practices in such a way as to improve the coherence of the whole set of existing arrangements. As our case study of water emphasized, measures taken to enhance agricultural production may lead to severe degradation of water quality, whatever their contribution to the achievement of food security. As a potential "adjudicatory norm," the *Grundnorm* could help to build cooperative relationships among the Sustainable Development Goals and their targets by treating them as instruments for achieving the same basic purpose, and therefore as elements of a common program (Kim 2016).

Developing countries may hesitate to endorse this idea for the same reason they are reluctant to accept any language that seems to downgrade the importance of economic development. But fear that a sustainability *Grundnorm* would serve to defend the privileges of the rich would be misplaced. The main threat to the earth's life-support systems comes from overconsumption by rich people, not from the struggle of poor people to satisfy basic human needs. A sustainability *Grundnorm* would recognize the right of all people to improved well-being, and the sustainability goals would serve to spell out important implications of that right. A sustainability *Grundnorm* would have implications for the allocation of resources among the members of any particular generation. But this is not the primary concern. Rather, the main purpose of articulating such a *Grundnorm* would be to highlight a precondition for human well-being at all times and in all places.

Conclusion

Although it is understandable that some approach the articulation of the Sustainable Development Goals largely as a matter of launching the second round of the Millennium Development Goals, the real challenge of framing the Sustainable Development Goals and implementing them effectively is more complex. Goal setting as a governance strategy in this context is a matter of responding, at one and the same time, to the legitimate concerns of those seeking to finish the job of fulfilling the Millennium Development Goals and of those seeking to address the emerging threats to human well-being associated with the onset of the Anthropocene and, in the process, to preserve a safe operating space for humanity. A common response to situations of this sort is to adopt an inclusive approach, one that includes in the list of proposed Sustainable Development Goals the concerns of all major constituencies. It also requires a "backcasting" approach to start, by drawing an ideal picture for the future. But this is not a recipe for success. As the outcome document from the 2012 UN Conference on Sustainable Development reminds us, the goals should be both "concise" and "limited in number" (UNGA 2012, par. 247). The ability of those responsible for crafting the Sustainable Development Goals to find a way to fulfill this requirement will go far toward determining the fate of this important effort to forge an innovative pathway in the realm of international governance. So will, in the next phase, the ability of the UN system and its member states to support these goals by establishing institutional arrangements and operational practices that combine the need for inspirational high-level leadership with the mobilization of important stakeholders and necessary resources at multiple levels (see Bernstein, chapter 9).

References

Bardach, Ann Louise. 2014. Lifestyles of the Rich and Parched. *Politico Magazine*. Available at: http://www.politico.com/magazine/story/2014/08/california-drought-lifestyles-of-the-rich-and-parched-110305.html.

Barrett, Scott. 2007. *Why Cooperate? The Incentive to Supply Global Public Goods*. Oxford: Oxford University Press.

Berkes, Fikret, Terry P. Hughes, Robert S. Steneck, James A. Wilson, David R. Bellwood, Beatrice Crona, Carl Folke, et al. 2006. Globalization: Roving Bandits and Marine Resources. *Science* , 311: 1557–1558.

Bosselmann, Klaus. 2008. *The Principle of Sustainability: Transforming Law and Governance*. Aldershot: Ashgate.

Bridgewater, Peter, Rakhyun E. Kim, and Klaus Bosselmann. 2015. Ecological Integrity: A Relevant Concept for International Environmental Law in the Anthropocene? *Yearbook of International Environmental Law*, 25: 61–78.

Chapin, F. Stuart, Gary P. Kofinas, and Carl Folke, eds. 2009. *Principles of Ecosystem Stewardship: Resilience-Based Natural Resource Management in a Changing World*. New York: Springer.

Cornell, Sarah E., I. Colin Prentice, Joanna I. House, and Catherine J. Downy. 2012. *Understanding the Earth System: Global Change Science for Application*. Cambridge, UK: Cambridge University Press.

Fagin, Dan. 2013. *Toms River: A Story of Science and Salvation*. New York: Bantam Books.

Fisher, Douglas. 2013. *Legal Reasoning in Environmental Law: A Study of Structure, Form and Language*. Cheltenham: Edward Elgar.

Fisher, Irving. 1930. *The Theory of Interest*. New York: Macmillan Co.

Gleick, Peter, Newsha Ajami, Juliet Christian-Smith, Heather Cooley, Kristina Donnelly, Julian Fulton, Mai-Lan Ha, et al. 2014. *Biennial Report on Freshwater Resources*. vol. 8. The World's Water. Washington, DC: Island Press.

Griggs, David, Mark Stafford-Smith, Owen Gaffney, Johan Rockström, Marcus C. Öhman, Priya Shyamsundar, Will Steffen, et al. 2013. Sustainable Development Goals for People and Planet. *Nature* 495: 305–307.

International Court of Justice. 1996. Legality of the Threat or Use of Nuclear Weapons. Advisory opinion. *ICJ Reports* 1996: 226–267.

International Union for Conservation of Nature Environmental Law Programme. 2015. *Draft International Covenant on Environment and Development.* Fifth edition. Gland: International Union for Conservation of Nature.

Kahneman, Daniel. 2011. *Thinking, Fast and Slow.* New York: Farrar, Straus and Giroux.

Kahneman, Daniel, and Amos Tversky. 1979. Prospect Theory: An Analysis of Decisions under Risk. *Econometrica* 47: 263–291.

Kanie, Norichika, Michele M. Betsill, Ruben Zondervan, Frank Biermann, and Oran R. Young. 2012. A Charter Moment: Restructuring Governance for Sustainability. *Public Administration and Development* 32: 292–304.

Kelsen, Hans. 1967. *Pure Theory of Law.* Berkeley: University of California Press.

Kim, Rakhyun E. 2016. In press. The Nexus between International Law and the Sustainable Development Goals. *Review of European Comparative and International Environmental Law.*

Kim, Rakhyun E., and Klaus Bosselmann. 2013. International Environmental Law in the Anthropocene: Towards a Purposive System of Multilateral Environmental Agreements. *Transnational Environmental Law* 2: 285–309.

Kim, Rakhyun E., and Klaus Bosselmann. 2015. Operationalizing Sustainable Development: Ecological Integrity as a *Grundnorm* of International Law. *Review of European, Comparative & International. Environmental Law (Northwestern School of Law)* 24: 194–208.

Kuo, Lily. 2014. China Has Launched the Largest Water-Pipeline Project in History. *The Atlantic,* March 7.

Kydland, Finn E., and Edward C. Prescott. 1977. Rules Rather than Discretion: The Inconsistency of Optimal Plans. *Journal of Political Economy* 85: 473–491.

Lenton, Timothy M., Hermann Held, Elmar Kriegler, Jim W. Hall, Wolfgang Lucht, Stefan Rahmstorf, and Hans-Joachim Schellnhuber. 2008. Tipping Elements in the Earth's Climate System. *Proceedings of the National Academy of Sciences of the United States of America* 105: 1786–1793.

Little, Jane Braxton. 2009. The Ogallala Aquifer: Saving a Vital U.S. Water Source. *Scientific American,* March 1.

Muys, Bart. 2013. Sustainable Development within Planetary Boundaries: A Functional Revision of the Definition Based on the Thermodynamics of Complex Social-Ecological Systems. *Challenges in Sustainability* 1: 41–52.

Open Working Group on Sustainable Development Goals. (August 2014). Proposal of the Open Working Group for Sustainable Development Goals. *UN Doc. A* 68/970: 12.

Philippines Supreme Court. 1994. *Minors Oposa v. Secretary of the Department of the Environment and Natural Resources*. 33 ILM 173 (1994).

Reisner, Marc. 1993. *Cadillac Desert: The American West and Its Disappearing Water*. New York: Penguin Books.

Rockström, Johan, Will Steffen, Kevin Noone, Åsa Persson, F. Stuart Chapin, III, Eric F. Lambin, Timothy M. Lenton, et al. 2009. A Safe Operating Space for Humanity. *Nature* 461: 472–475.

Sachs, Jeffrey D. 2015. *The Age of Sustainable Development*. New York: Columbia University Press.

Scanlon, Bridget R., Ian Jolly, Marios Sophocleous, and Lu Zhang. 2007. Global Impacts of Conversions from Natural to Agricultural Ecosystems on Water Resources: Quantity versus Quality. *Water Resources Research* 43. doi: 10.1029/2006WR005486 .2007.

Schelling, Thomas C. 1995. Intergenerational Discounting. *Energy Policy* 23: 395–401.

Science and Environmental Health Network and the International Human Rights Clinic at Harvard Law School. 2008. Models for Protecting the Environment for Future Generations. Available at: http://www.sehn.org/pdf/Models_for_Protecting _the_Environment_for_Future_Generations.pdf.

Seoul Administrative Court. 2001. Judgment 2000Gu12811 (known as the "future generations' lawsuit").

Steffen, Will, Åsa Persson, Lisa Deutsch, Jan Zalasiewicz, Mark Williams, Katherine Richardson, Carole Crumley, et al. 2011. The Anthropocene: From Global Change to Planetary Stewardship. *Ambio* 40: 739–761.

Steffen, Will, Katherine Richardson, Johan Rockström, Sarah E. Cornell, Ingo Fetzer, Elena M. Bennett, Reinette Biggs, et al. 2015. Planetary Boundaries: Guiding Human Development on a Changing Planet. *Science* 347: 1259855.

Steffen, Will, Angelina Sanderson, Peter Tyson, Jill Jäger, Pamela A. Matson, Berrien Moore, III, Frank Oldfield, et al. 2004. *Global Change and the Earth System: A Planet Under Pressure*. Berlin: Springer.

Stern, Nicholas. 2007. *The Economics of Climate Change: The Stern Review*. Cambridge, UK: Cambridge University Press.

Supreme Court of Montana. 1999. *Environmental Information Center v. Department of Environmental Quality*, 988 P.2d 1236.

Underdal, Arild. 2010. Complexity and Challenges of Long-term Environmental Governance. *Global Environmental Change* 20: 386–393.

UNGA, United Nations General Assembly. 2012. The Future We Want. UN Doc. A/RES/66/288.

von Weizsäcker, Ernst Ulrich, Oran R. Young, and Matthias Finger, eds. 2005. *Limits to Privatization: How to Avoid Too Much of a Good Thing*. London: Earthscan.

World Commission on Environment and Development. 1987. *Our Common Future*. Oxford: Oxford University Press.

Young, Oran R. 2013. *On Environmental Governance: Sustainability, Efficiency, and Equity*. Boulder, CO: Paradigm Publishers.

4 Global Goal Setting for Improving National Governance and Policy

Frank Biermann, Casey Stevens, Steven Bernstein, Aarti Gupta, Norichika Kanie, Måns Nilsson, and Michelle Scobie

Can better governance, in itself, be a subject for global goal setting? This question stands at the center of this chapter, which focuses on the inclusion of "governance goals" in global goal-setting mechanisms, especially the Sustainable Development Goals agreed upon by the UN General Assembly in September 2015 as part of its 2030 Agenda for Sustainable Development (UNGA 2015). While our discussion is inspired by the negotiations around governance goals and targets within the context of Sustainable Development Goals, we seek to build a broader analytic approach that goes beyond the integration of governance in this specific context.

We define governance here as the purposeful and authoritative steering of societal processes by political actors. Governance thus includes traditional activities by governmental actors, such as laws, policies, and regulations; planning practices, rule systems, and procedures at subnational levels; and certain actions by nongovernmental actors, such as standards set by civil society networks or public-private partnerships, as long as these activities include a claim to authority, have some legitimacy, and are designed to steer behavior. While there are some debates in the literature regarding the exact boundaries of what counts as governance, there is general agreement that authority and steering are its two core components (Rosenau 1995; Bernstein 2010; Biermann 2014). In addition, the identification of issues, agenda setting, information gathering and processing, negotiation, setting policy goals, and their implementation and monitoring are all part of governance. Finally, globally defined goals, such as the Sustainable Development Goals, can themselves be powerful governance tools with a major impact on the behavior of governments, international organizations, and nongovernmental actors. This aspect is addressed in two related chapters in this book (see Young, this volume, chapter 2; Young et al., this volume, chapter 3).

We focus in this chapter on three core qualities of governance, which we analyze in more detail: good governance, effective governance, and equitable governance. In our understanding, "good" governance focuses on *qualitative characteristics of governance* such as accountability, transparency, participation, and the rule of law. Effective governance looks at improving the *overall problem-solving capacity* of governance. Equitable governance focuses on the *processes and distributive outcomes* of governance, including the need to protect the interests of poor and vulnerable groups.

The second section of the chapter discusses the emergence of governance on the international development agenda and its inclusion in the Sustainable Development Goals. We also discuss how particular framings of governance have had a significant impact on the indicators available, and on integrating governance into sustainable development. The third section explores the three different dimensions of governance and their application to sustainable development. The fourth section then reflects upon advantages and disadvantages for pursuing governance as a stand-alone goal or as integrated into issue-specific goals and objectives within international institutions and agreements.

Governance on the International Agenda

Governance emerged on the international agenda in the 1990s, fueled to a large extent by the World Bank and the International Monetary Fund (Sundaram and Chowdhury 2012). In combining institutional economic theory with new public management and a Weberian view of the state, the operationalization of governance by these organizations was largely technocratic in practice (Andrews 2008). Their focus, furthermore, has been on the "good governance" dimension—on issues such as corruption, transparency, accountability, and the rule of law—with less attention to either "effective" or "equitable" governance. Also, the attention of these organizations has been more on the process of governance and not so much on the medium- and long-term results or outcomes of governance. In practice, international financial institutions mandated experts to design governance measures, and then used such measures to first target aid and eventually to make aid conditional upon "good" governance indicators (Best 2014a). While encouraging transparency, accountability, and the involvement of civil society within countries, such ideals did not necessarily translate into the practices of international financial institutions themselves (Woods 2000).

Earlier discussions had been largely technocratic and focused on development of complex indicators, centered on issues such as limiting corruption and supporting the rule of law, as well as enhancing transparency, accountability, and participation. In the 1990s, this focus shifted to include the complex political relationships that support governance institutions (Leftwich 1994; Sundaram and Chowdhury 2012). This signaled a shift from a supply-side focus on governance that tries to incentivize particular institutions to improve their governance, with a demand-side focus on supporting actors in domestic and transnational contexts to demand better governance (Best 2014b). Although a wider range of actors are now debating indicators of good governance, the World Bank and the International Monetary Fund are still at the center of much of this work. As a result of these debates, these actors have adjusted their approach from a one-size-fits-all model of governance that animated much of the earlier phases in favor of a more nuanced approach, emphasized in the 2010 Seoul Development Consensus for Shared Growth (Stiglitz 2008).

While the hypothesis that good governance will facilitate economic growth is considered a truism in much of the policy community, its veracity has not been conclusively shown (Sundaram and Chowdhury 2012). Some countries that have achieved significant economic growth since 1945 have done so in authoritarian and corrupt contexts (Wilkin 2011). Similarly, the liberal model of governance from Western Europe and the United States has proven neither as monolithic nor as unproblematic as the hypothesis is interpreted to be. Furthermore, it may well be that economic growth fosters good governance, rather than the other way around (Chang 2011).

Multiple studies on this relationship have shown that "countries have only improved governance with development, and that good governance is not a necessary precondition for development" (Sundaram and Chowdhury 2012, 9; also Holmberg and Rothstein 2011). Kwon and Kim argue, for instance, that the "empirical evidence does not support the hypothesis that good governance leads to poverty reduction. Good governance alleviates poverty only in middle-income countries, not in least developed ones" (Kwon and Kim 2014). Thus, while it is likely that institutional breakdown has a clear relationship to economic poverty, this does not mean that the relationship holds across all levels or contexts of governance or development (Wilkin 2011; Khan 2007; Aron 2000). Even the governance agenda around limiting corruption has faced challenges on being institutionalized and has produced uneven results. Some empirical studies in developing countries have not shown corruption to be a drag on development (Meon and Weill 2010), and some anti-corruption programs have been taken over by political elites (Mungiu-Pippidi 2006). Other studies, however, indicate that "corruption has a negative effect on per-capita GDP growth" (Ugur 2014). All of this is notwithstanding, of course, that there are important normative reasons that speak for continuous efforts in anti-corruption policies, including considerations of fairness, legitimacy, and trust in government.

As for sustainability more narrowly defined, the governance agenda has included a focus on sustainability in developing countries from its beginning, often linked to the World Bank publication *From Crisis to Sustainable Growth: Sub-Saharan Africa: A Long-term Perspective Study* (World Bank 1989; Van Doeveren 2011). However, in this report, and in much of the discussion that followed, the international economic institutions focused the discussion primarily on governance for sustained economic growth, while environmental or social sustainability did not figure prominently.

In addition, most attention has been placed so far at the national level of governance. While much scholarship has analyzed the relevance of better governance for the success of national policies, from welfare to environmental protection, the debate on *international* standards for improved governance is of more recent origin (Hulme, Savoia, and Sen 2015). Even though concepts such as "good governance" have long been central to the governance debate in development cooperation, detailed standards of such "good governance" have been limited so far in international agreements and institutions. For example, the Millennium Development Goals, adopted in 2000, did not include explicit goals on governance. Some have suggested that this absence has significantly limited efforts to reduce poverty worldwide (Sachs and McArthur 2005) and narrowed the focus from the broader Millennium Declaration, which included attention to equity, human rights, and other elements of the broader governance agenda (Browne 2014; Fukuda-Parr, Yamin, and Greenstein 2014).

Governments began to address this concern in negotiations of the Sustainable Development Goals. At the 2012 UN Conference on Sustainable Development, governments agreed that "good governance and the rule of law at the national and international levels are essential for sustained, inclusive and equitable economic growth, sustainable development and the eradication of poverty and hunger." Similarly, the 2014 UN Development Programme report argued that the quality of governance plays "a defining role in supporting the pillars" of the Sustainable Development Goals (UNDP 2014). Likewise, a comprehensive Foresight Process organized by the UN Environment Programme identified "aligning governance to the challenges of global sustainability" as the most urgent emerging issue related to the global environment (UNEP 2012).

Calls for specific governance goals at the international level can also be found among a wide range of nongovernmental organizations and individuals. The UN High-level Panel of Eminent Persons argued that governance should be included as part of the Sustainable Development Goals, which would help accomplish "a fundamental shift—to recognize peace and good governance as core elements of wellbeing, not optional extras" (United Nations 2013). The "Action Agenda" advanced by the Sustainable Development Solutions Network (SDSN 2013) made a similar call, as did a 2014 statement by 50 civil society organizations (Transparency and Accountability Initiative 2014). Other organizations have called for integrating governance into issue-specific goals relating, for example, to water, food, or gender (UNDESA 2014).

Yet others have sought to link progress in governance at the national level with similar transformations of institutions of global governance. For example, the International Development Law Organization (2014) recommended that the "rule of law [widely considered part of the good governance agenda] is also relevant at the global level to the legal and institutional frameworks for trade, investment, intellectual property, technology transfer and addressing climate change—where fairer rules would create a more equitable, inclusive and sustainable model of development." Conceptually, this discussion reaches beyond the limited confines of the Sustainable Development Goals, as goals for better governance could theoretically be integrated into numerous international institutions and agreements, including most multilateral environmental agreements.

Following these various inputs, governments supported governance as one of 11 thematic areas for consultation around which the Sustainable Development Goals were negotiated. The 2030 Agenda for Sustainable Development (UNGA 2015) now states in paragraph 9, "We envisage a world ... in which democracy, good governance and the rule of law as well as an enabling environment at national and international levels, are essential for sustainable development, including sustained and inclusive economic growth, social development, environmental protection and the eradication of poverty and hunger." In addition, while not using the term "governance" explicitly, Sustainable Development Goal 16 calls on states to "promote peaceful and inclusive societies for sustainable development, provide access to justice for all and build effective, accountable and inclusive institutions at all levels." The targets of this goal include issues such as strengthening the rule of law, decreasing corruption, enhancing participation and accountability of institutions, transparency of decision making, and so forth. Some of the other specific goals also called for governance integration, for example with the target to develop integrated water resource management at all levels under Goal 6 about freshwater availability.

The specific ways in which the global governance agenda has developed show the need to emphasize the three different dimensions of governance, as conceptualized in this chapter. While there is an assumption in the phrasing around most governance goal proposals that achieving "good governance" will lead to "effective" and "equitable" governance, the operationalization of governance makes the relationship less clear. Existing indicators and indexes of governance neither operationalize nor clarify such potential relationships (Thomas 2010; Gisselquist 2014). For example, in Central America, while the governance agenda has focused on corruption in the government, it has been largely silent on private accumulation of capital and exacerbation of inequalities (Horton 2012). A causal relationship between one dimension of governance and another is thus not empirically supported and needs analysis on its own terms.

Good Governance

As our discussion has shown, of all dimensions of governance the quest for "good governance" has received the most attention, including within it issues as diverse as participation, transparency, accountability, public access to information, combating corruption, human rights, and strengthening the rule of law. Each of these aspects, however, raises thorny questions about the benefits of a "one-size-fits-all" approach that typically follows the political systems of industrialized countries, versus approaches that take account of different political conditions (Overseas Development Institute 2013; Overseas Development Institute 2014).

Given the broad scope of good governance, there are numerous indicators that seek to measure the "quality of governance" at the national level and draw on various combinations of indexing, expert coding, and perception surveys (see also table 4.1).[1]

Indexes for good governance, for instance, often collate multiple measures to generate overall governance scores for countries. Examples include the Worldwide Governance Indicators (World Bank 2014); the Ibrahim Index of African Governance (Mo Ibrahim Foundation 2014); the Government Effectiveness Indicator (Millennium Challenge Corporation 2014);

Table 4.1
Indicators of Governance

Governance indicator	Organization	Type of organization	Primary method for indicator	Problems
Corruption Perception Index	Transparency International	Nongovernmental organization	Perception surveys	Survey may not capture all relevant stakeholder opinions
Democracy Index	Economist Intelligence Unit	For profit	Expert assessment	For-profit company
Economics and Country Risk	IHS	For profit	Expert assessment	For-profit company
Freedom in the World	Freedom House	Nongovernmental organization	Expert assessment	
Gallup World Poll	Gallup, Inc.	For profit	Perception surveys	Random sampling may not account for excluded minorities or other inequalities
Global Barometer Surveys	Secretariat Latinobarometer	Nongovernmental organization	Perception surveys	Regional efforts vary widely
Global Right to Information Rating	Right to Information Europe and Center for Law and Democracy	Nongovernmental organization	Expert assessment	Based only on legislation and no other limitations on freedom to information
Governance Effectiveness Indicator	Millennium Challenge Corporation	United States government	Index	Tied to US development agenda

Table 4.1 (continued)

Governance indicator	Organization	Type of organization	Primary method for indicator	Problems
Ibrahim Index of African Governance	Mo Ibrahim Foundation	Nongovernmental organization	Index	Links to African civil society unclear
Rule of Law Index	World Justice Project	Nongovernmental organization	Expert assessment	
Social Institutions and Gender Index (SIGI)	Organisation for Economic Co-operation and Development	International organization	Index	May use indicators that focus more on OECD country experiences
Transformation Index	Bertelsmann Stiftung	Nongovernmental organization	Expert assessment	
UN Rule of Law Indicators	United Nations	International organization	Expert assessment	Robust application is unclear
World Governance Indicators	World Bank	International organization	Index	Tied closely to international financial institutions "good governance" agenda of the 2000s
World Values Survey	World Values Survey Association	Nongovernmental organization	Perception Surveys	Random sampling may not account for excluded minorities or other inequalities

and the Social Institutions and Gender Index (OECD 2014b); see generally Gisselquist (2014).

A second way of measuring the quality of governance at the national level is to invite global or local experts to fill in questionnaires and then quantify the results. Examples are the Democracy Index (Economist Intelligence Unit 2014); the Freedom in the World index (Freedom House 2014); the Economics and Country Risk (formerly Global Insight) index (IHS 2014); the Rule of Law Index (World Justice Project 2014); the Global Right to Information Rating (Right to Information Europe and Center for Law and Democracy 2014); the Transformation Index (Bertelsmann Stiftung 2014); the Quality of Government data set (Teorell, Dahlström, and Dahlberg 2011), and the UN Rule of Law Indicators (United Nations 2014).

A third method involves gathering public survey data focused on perceptions of specific issues by select members of a population or sampling randomly from the general population. Examples are the Gallup World Poll (Gallup, Inc. 2014); the World Values Survey (World Values Survey Association 2014); the Global Barometer Surveys (Global Barometer 2014); and the Corruption Perception Index (Transparency International 2014).

A fourth form of developing country-relevant indicators for good governance is the use of country peer-review mechanisms. Although these do not develop cross-country indicators of governance, the mechanism can build on, and go beyond, the indicators above. Examples are the African Peer Review Mechanism (African Union 2014) and the peer-review mechanisms by the Organisation for Economic Co-operation and Development's Development Assistance Committee (OECD 2014a).

As noted above, in many respects the lead organizations in building rigorous indicators of governance were the international financial institutions. Their indicators often emphasized protection of property rights, anti-corruption, and rule of law as crucial aspects of good governance. The Worldwide Governance Indicators by the World Bank are the most recognized indexes on governance, collecting, at the time of this writing, data from 32 different sources to construct elaborate multi-indicator scores of six different components of governance (Kaufmann, Kraay, and Mastruzzi 2010). The index relies heavily on elite perception surveys and largely ignores connections between different indicators (Kurtz and Schrank 2012; Thomas 2010; Andrews 2008), and has not remained uncontested.

However, the Worldwide Governance Indicators are not the only measure of good governance. Others are compiled by activist organizations, think tanks, economic research organizations, or other international organizations and, as a result, often carry implicitly or explicitly the bias of their respective organizations, which may reduce their legitimacy in some countries and contexts. However, the methodology and experiences of these measures can still prove useful for generally assessing progress toward better governance. An example is the overlapping system of indicators by the "g7+" alliance of fragile states that includes a few universal indicators all participating countries agree upon, complemented with a menu of indicators that allows countries to select some that they find most appropriate and complement these with others developed in their own national context (g7+ 2012). In addition, there are opportunities for creative use of proxy indicators and the creation of new indicators to include additional dimensions of good governance (Andrews 2008).

Most targets and indicators of good governance, however, are subject to political contestation. While some targets such as ending corruption or combating money laundering may find general agreement, the indicators for these targets may be contested. Developing countries, in particular, are sensitive to the potential for "good governance" targets and indicators to be used in (re)designing programs on official development assistance, technology transfer, or trade benefits. Targets that focus on the overall efforts in implementing agreements could help bridge this disagreement. As one example, a 2014 civil society proposal emphasized targets on transparency and freedom of information in projects on sustainable development (Transparency and Accountability Initiative 2014). However, such reporting requirements might also be onerous for developing countries with limited technical and human resources. Thus, facilitating governance for monitoring and reporting becomes important to build transparency and improve accountability for developing countries. Conceptually, then, the issue is to identify the development deficits in each case (at the level of country, organization, or community) and to identify what governance targets will be most relevant to achieve increased development in light of the social, economic, and environmental realities (such as limited reporting capacity, corruption, low productivity, competitiveness, and so forth).

Sustainable Development Goal 16, as it was finally agreed upon in September 2015, now combines different aspects of "good governance" as part of the overall goal to "promote peaceful and inclusive societies for sustainable development, provide access to justice for all and build effective, accountable and inclusive institutions at all levels" (UNGA 2015). Targets include elements of the "good governance" agenda, such as to "promote the rule of law at the national and international levels and ensure equal access to justice for all" (16.3); "substantially reduce corruption and bribery in all their forms" (16.5); "develop effective, accountable and transparent institutions at all levels" (16.6); "ensure responsive, inclusive, participatory and representative decision making at all levels" (16.7); or "ensure public access to information and protect fundamental freedoms, in accordance with national legislation and international agreements" (16.10).

In integrating elements of "good governance" explicitly, the Sustainable Development Goals thus go substantially beyond the Millennium Development Goals, even though the formulations remain broad and qualified by reference to national legislation. Yet overall, this broadness might also foster, in the coming years, international agreement and build alliances among different countries with different priorities, capacities, and experiences, with novel opportunities to link different aspects of good governance in goal-framing exercises that can bring countries with different interests together.

Finally, "good governance" is a matter that concerns not only governments and intergovernmental organizations, but also nonstate actors. Given the proliferation of partnerships, action networks, and transnational governance arrangements involved in sustainable development, accountability mechanisms that incorporate good governance criteria might also be appropriately applied to them. As Bernstein (this volume, chapter 9) suggests, the Sustainable Development Goals—and governance processes in support of them within the United Nations—might encourage the application of the United Nations' voluntary accountability framework to any partnership or transnational governance arrangement involved in sustainable development. Such a focus resonates with findings on the importance of good internal governance of public-private partnerships for their effectiveness and legitimacy (Biermann et al. 2007; Bäckstrand et al. 2012; Bäckstrand and Kylsäter 2014).

Effective Governance

With regard to effective governance, the question arises whether governments can in fact agree on a set of measures to assess the capacity of governance systems to effectively address today's complex sustainability challenges. In the context of global environmental change and unsustainable development pathways, it is particularly important to strengthen the institutional basis for long-term decision making and for integrated implementation of sustainable development policies (Nilsson and Persson 2012;

Kanie et al. 2014). Governments need not agree on a specific institutional basis for long-term decision making and policy integration, but rather on the need to enhance capacities of existing institutions at local, subnational, national, or regional levels, and even of authoritative governance arrangements that may operate in nontraditional governance spaces such as partnerships or transnational governance mechanisms or initiatives.

In this context, the process leading up to the Sustainable Development Goals tended either to define effective governance as "means of implementation" (focused on official development assistance and technology transfer, though later extended significantly into domestic resource mobilization, remittances, private finance, and technical capacity, among other means and mechanisms) or to focus on issue-specific discussions on effective institutions. The focus on capacity building for integrated and long-term policy making for achieving the Sustainable Development Goals has been limited. Furthermore, reaching political agreement on means of implementation was complex, and targets that focus on national goal setting and long-term policy outlooks appeared more politically tractable than targets for means of implementation.

Existing indicators of effective governance are also largely limited to specific issues, such as water governance. Some possibilities for wider integration include expanded and more systemic stress testing of national institutions (as used by the World Bank, the International Monetary Fund, and the Group of 20 major economies as regards central banks); proxy measures linked to governance decisions (for example, the under-age-5 mortality rate, see Andrews, Hay, and Myers 2010); and surveys on people's perception of effective governance.

In the Sustainable Development Goals, effective governance has eventually been integrated into a number of issue-specific goals. Of the 17 Sustainable Development Goals, nine have targets specifying objectives for improving management of resources or social problems. For example, Goal 4 on education states, "By 2030, ensure that all learners acquire the knowledge and skills needed to promote sustainable development, including, among others, through education for sustainable development and sustainable lifestyles, human rights, gender equality, promotion of a culture of peace and non-violence, global citizenship and appreciation of cultural diversity and of culture's contribution to sustainable development" (4.7). Goal 6 on water and sanitation includes a specification for "implementing integrated water resources management at all levels" (6.5) and "the participation of local communities in improving water and sanitation management" (6b) for integrating local communities into sanitation planning.

While implementation will determine the impact of these aspects, they provide openings for more effective governance. In addition, the governance-related Goal 16 explicitly addresses the effectiveness of governance, for instance in its call to "develop effective, accountable and transparent institutions at all levels" (16.6) and to "strengthen relevant national institutions, including through international cooperation, for building capacity at all levels, in particular in developing countries, to prevent violence and combat terrorism and crime" (16.a).

In sum, effective governance includes an emphasis on strengthened capacities for long-term planning, going beyond means of implementation. Corresponding goals and targets for effective governance are often included within the issue-specific Sustainable Development Goals. Development of national sustainable development plans; statistical and other related monitoring; data and analytic capacity; and human resources for governance and policy-making capacity are among the concrete needs for effective governance that ought to be prioritized.

Equitable Governance

Finally, discussions of governance must not be separate from considerations of equity. Governance plays a crucial role in the distribution of outcomes, hence striving for equitable or fair solutions to public policy problems seems critical. The notion of "no one left behind" represents the importance of equitable governance in the overall context of the 2030 Agenda for Sustainable Development (UNGA 2015).

Numerous indicators can allow equitable governance issues to be integrated into governance goals within international institutions. The Gini index that seeks to represent the income or wealth distribution within a country or community is one of the most prominent. The index—which scales wealth distribution with a value of zero indicating perfect wealth equality between all citizens and a value of one indicating maximum inequality—can be applied to a variety of assets (including wealth, income, education, and others) and allows for measurement over time. While targets for Gini index outcomes could be a part of governance goals in international institutions, domestic income distribution remains strongly entangled in value differences between countries, which will make any operationalization of equality through quantitative agreements on reducing the Gini coefficient difficult. Also, inequality of income and access to decision-making power varies significantly across countries, as issues of wealth inequality interact with minority rights, women's rights, and other specific national contexts.

Some have argued, as a result, that indicators that can be disaggregated into different social categories should have constituted the primary means for including equity concerns into global governance goals (Sustainable Development Solutions Network 2015). Yet few existing good governance or effective governance indicators have such levels of disaggregation. This may provide an opportunity to refine dashboards of various indicators to include equity in processes and outcomes of governance, in addition to good governance and effective governance. Another alternative is the agreement by governments on qualitative propositions that state the undesirability of high levels of domestic inequality as a key element of global progress toward inclusive sustainable development, leaving the exact target values undefined.

In addition, equitable outcomes can be operationalized in the form of *absolute targets*, notably in the agreement on special protection awarded to the poorest people in a community or a country (Doyle and Stiglitz 2014). The global agreement on eradication of poverty and reduction in hunger and malnutrition are examples from the Millennium Development Goals that stipulate a normative statement about minimum thresholds in national wealth or income distribution that are not acceptable and thus require urgent political action. In the negotiations of the Sustainable Development Goals, governments discussed, among other things: inequalities among social groups, inequalities between social groups, empowerment of marginalized groups, pursuit of gender equality, strengthening of social protection systems, promoting higher income growth among the poorest segments of societies, and inequalities in opportunities and economic chances, especially of women and girls (Open Working Group 2014).

The Sustainable Development Goals now include a number of Goals and targets that address distributive outcomes of governance. For example, Goal 10, "Reduce inequality within and among countries" includes targets that focus on reducing high levels of inequality in the distribution of wealth or income within and among countries, and Goal 5, "Achieve gender equality and empower all women and girls," requires that women be granted equal opportunities and rights to economic and technological resources. However, adequate indicators for all targets have yet to be worked out, and progress in the combination of different goals and targets on equality should be carefully measured, such as access to education of women among the poorest populations. The question in the next period of implementing the Sustainable Development Goals is thus how and whether these goals and targets can be translated into specific institutional settings that require governments to design policies that reduce high levels of inequality and that further the eradication of poverty. It remains also an open question to what extent the broad promise of Goal 10 to reduce inequality also "among countries" will materialize in the coming decade (Edwards and Romero 2014).

Conceptualizing Governance as a Stand-alone or as an Integrated Goal

These three dimensions of governance provide a useful point of departure for considering the advantages and disadvantages of pursuing governance either as a standalone "governance goal" in international institutions and agreements, or integrating it into issue-specific goals. The different dimensions of governance will manifest themselves in distinct ways because of different political alignments, availability of indicators, and mobilization possibilities. This section thus reflects on the pros and cons of either standalone governance goals or the cross-cutting integration of governance concerns in issue-specific goals, both of which are part of the 2030 Agenda for Sustainable Development.

Two sections in the Sustainable Development Goals focus on governance in a standalone fashion, yet both leave significant dimensions of governance unspecified. Goal 16 to "Promote peaceful and inclusive societies for sustainable development, provide access to justice for all and build effective, accountable and inclusive institutions at all levels" highlights general aspects of good governance, but somewhat ignores effective governance and equitable aspects of governance. Section 16.3 calls for promotion of the rule of law to "ensure equal access to justice for all" and 16.6 calls for states to "Develop effective, accountable and transparent institutions at all levels." Removing barriers to equality, rather than any positive promotion of equality, is the main focus of sections 16.9 that "By 2030, provide legal identity for all, including birth registration," and 16.b, "Promote and enforce non-discriminatory laws and policies for sustainable development." The focus on good governance dimensions like removing corruption, improving the rule of law, and increasing transparency, accountability, and participation does not include similar sections for focusing on coherence of policy-making capacities or equitable governance aspects.

Sustainable Development Goal 17 now includes three targets on "Policy and institutional coherence" (17.13–17.15), even though this poses significant measurement challenges as well. Although there are a number of proxies for issues such as capacities for effective governance that may serve specific needs, there still are no universally accepted measures of state administrative and legal capabilities (Hulme, Savoia, and Sen 2015). Second, capacity may mean different things in different countries and political contexts. A potential avenue for integrating effective governance into a standalone goal could have been through governance assessment indicators (for example, an indicator to assess problem solving capacity). This was the procedure followed, for example, by Mongolia with its initiative for a Millennium Development Goal on democratic governance, which started with an assessment procedure by which the quality of their democracy was openly assessed. Third, there is also a risk that comprehensive standalone goals on governance downplay the equity aspects of governance outcomes, especially when it comes to the development of indicators.

A parallel approach in the 2030 Agenda for Sustainable Development has been the integration of governance-related dimensions into issue-specific goals, such as on water and sanitation (Goal 6), health and wellbeing (Goal 3), or gender equality (Goal 5). This approach creates different challenges. One challenge is that such an approach might support a focus on the easiest or most politically palatable aspects of governance, while leaving out the more difficult or contested dimensions of governance, for instance those related to long-term environmental sustainability and equity of outcomes. This is further complicated by the fact that such integration is further advanced in certain issue-areas (for example, water and sanitation) than in others (for example, education). The challenge of incorporating all three dimensions of governance is evident in the 2030 Agenda for Sustainable Development, where good governance aspects are highlighted more significantly in some areas—such as water and sanitation—than in others, such as poverty eradication, energy access, or health.

Including effective governance in issue-specific goals also creates potential overlaps with the "means of implementation" and the "global partnership for sustainable development," both included in Goal 17. These household terms in UN processes are intentionally vague for purposes of political feasibility (and in order to reach agreement), but have been known to create confusion among national planners and policy makers.

Furthermore, assessing progress on effective governance requires a significant increase in the capacity of countries to measure, aggregate, report, and assess the results. Such components of effective governance could be integrated into the other globally agreed goals by building task-specific capacity (Pintér, Kok, and Almassy, this volume, chapter 5). However, there may not be consistent integration of this aspect across issues, and qualitative indicators (quality of education, nutrition, and so forth) may receive less attention. Also, it is important to distinguish between effective *processes* that enable the monitoring process and collection of data through capacity building or financing the staff needed for this activity, and *outcomes*, that is, the quality of data available for decision making.

Some aspects of equitable governance have also been integrated in issue-specific Sustainable Development Goals. Gender disaggregation of indicators, for example, has been promoted repeatedly and should be a core component of a governance agenda across goals and targets, not only in Goal 5. Some areas with existing inequality indicators might also see productive integration. One obvious example would be the necessity of linking indicators for targets in Goal 10 (on reducing inequality within and among countries) with those for Goal 17 (means of implementation and renewing the partnership for sustainable development). Both Goals identify similar mechanisms—aid, trade, financial flows, taxation, and so forth—but Goal 17 can only be achieved equitably if the targets in Goal 10, which specifically focus on targeting countries and parts of society where need is greatest, are taken seriously. As the above example suggests, the gap between goals with attention to governance concerns and those with insufficient attention is likely to be largest on equitable governance.

Conclusion

In sum, the inclusion of "governance goals" in international institutions and agreements is important to operationalize sustainable development. The Sustainable Development Goals include goals and targets that address the quality of governance, in terms of advancement toward good governance; the problem-solving effectiveness of governance, in the form of strengthening the governance capacities at national and local levels; and the distributive outcomes of governance, notably in terms of reducing extreme forms of inequality and advancing the interests of the poor. Even though the Goals include elements of all three dimensions, the way they are treated is quite different. Whereas "good" governance aspects are mostly confined to Goal 16, effective governance appears both across issue-specific Goals and in Goal 16, and equitable governance is treated in specific Sustainable Development Goals, as well as in Goal 10.

The uncertain relationship between good, effective, and equitable governance shows the importance of monitoring indicators that are related to these concepts at various levels of aggregation, both in relation to particular goals and national or even local contexts. In this regard, it is evident that governance must be recognized as important subject of review in its own right in order to generate shared experiences around the relationships and requirements of the three aspects of governance, and inputs into capacity and technical needs for improving governance for sustainable development. Including such analyses as regular components of the Global Sustainable Development Report could facilitate such learning, that is, through promoting the scientific study of how various governance arrangements, elements, and combinations of governance capacities, qualities, and techniques support progress on sustainable development. Similarly, governance must be a regular agenda item at the High-level Political Forum on Sustainable Development and other forums (global, regional, or country-level) where review and learning mechanisms operate as part of the follow-up on the Sustainable Development Goals. Finally, the "leaving no one behind" ethos of the Sustainable Development Goals requires good and effective governance in implementing each goal. In sum, policy sciences and governance studies in their own right ought to be included as an important component of the science-policy interface designed to support the achievement of the Sustainable Development Goals.

Note

1. We are grateful to Vincent Santos for assisting in the compilation of the data in this section.

References

African Union. 2014. African Peer Review Mechanism. Available at http://aprm-au.org/.

Andrews, Matt. 2008. The Good Governance Agenda: Beyond Indicators without Theory. *Oxford Development Studies* 36: 379–409.

Andrews, Matt, Roger Hay, and Jarrett Myers. 2010. Governance Indicators Can Make Sense: Under-five Mortality Rates are an Example. Research Working Paper Series 10–015. Available at http://research.hks.harvard.edu/publications/getFile.aspx?Id=541. Accessed April 1, 2014.

Aron, Janine. 2000. Growth and Institutions: A Review of the Evidence. *World Bank Research Observer* 15: 99–135.

Bäckstrand, Karin, Sabine Campe, Sander Chan, Ayşem Mert, and Marco Schäferhoff. 2012. Transnational Public-Private Partnerships. In *Global Environmental Governance Reconsidered*, ed. Frank Biermann and Philipp Pattberg, 123–147. Cambridge, MA: MIT Press.

Bäckstrand, Karin, and Mikael Kylsäter. 2014. Old Wine in New Bottles? The Legitimation and Delegitimation of UN Public-Private Partnerships for Sustainable Development from the Johannesburg Summit to the Rio+20 Summit. *Globalizations* 11 (3): 331–347.

Bernstein, Steven. 2010. When is Non-State Global Governance Really Governance? *Utah Law Review* 2010 (1): 91–113.

Bertelsmann Stiftung. 2014. Transformation Index. Available at http://www.bti-project.org/.

Best, Jacqueline. 2014a. *Governance Failure: Provisional Expertise and the Transformation of Global Development Finance*. New York: Cambridge University Press.

Best, Jacqueline. 2014b. The 'Demand Side' of Good Governance: The Return of the Public in World Bank Policy. In *The Return of the Public in Global Governance*, ed. Jacqueline Best and Alexandra Gheciu, 97–119. New York, NY: Cambridge University Press.

Biermann, Frank. 2014. *Earth System Governance: World Politics in the Anthropocene*. Cambridge, MA: MIT Press.

Biermann, Frank, Man-san Chan, Ayşem Mert, and Philipp Pattberg. 2007. Multistakeholder Partnerships for Sustainable Development: Does the Promise Hold? In *Partnerships, Governance and Sustainable Development. Reflections on Theory and Practice*, ed. Pieter Glasbergen, Frank Biermann and Arthur P. J. Mol, 239–260. Cheltenham: Edward Elgar.

Browne, Stephen. 2014. A Changing World: Is the UN Development System Ready? *Third World Quarterly* 35 (10): 1845–1859.

Chang, Ha-Joon. 2011. Institutions and Economic Development: Theory, Policy and History. *Journal of Institutional Economics* 7: 473–498.

Doyle, Michael D., and Joseph E. Stiglitz. 2014. Eliminating Extreme Inequality: A Sustainable Development Goal, 2015–2030. *Ethics & International Affairs* 28 (1): 5–13.

Economist Intelligence Unit. 2014. Democracy Index. Available at http://www.eiu.com/public/thankyou_download.aspx?activity=downloadandcampaignid=DemocracyIndex12.

Edwards, Martin S., and Sthelyn Romero. 2014. Governance and the Sustainable Development Goals: Changing the Game or More of the Same? *SAIS Review (Paul H. Nitze School of Advanced International Studies)* 34 (2): 141–150.

Freedom House. 2014. Freedom in the World. Available at http://www.freedomhouse.org/report-types/freedom-world.

Fukuda-Parr, Sakiko, Alicia Ely Yamin, and Joshua Greenstein. 2014. The Power of Numbers: A Critical Review of Millennium Development Goal Targets for Human Development and Human Rights. *Journal of Human Development and Capabilities* 15 (2–3): 105–117.

g7+. 2012. International Dialogue on Peacebuilding and State-building Working Group on Indicators. Available at: http://www.newdeal4peace.org/wp-content/uploads/2012/12/progress-report-on-fa-and-indicators-en.pdf.

Gallup, Inc. 2014. Gallup World Poll. Available at http://www.gallup.com/strategicconsulting/en-us/worldpoll.aspx?ref=f.

Gisselquist, Rachel M. 2014. Developing and Evaluating Governance Indexes: 10 Questions. *Policy Studies* 35 (5): 513–531.

Global Barometer. 2014. Global Barometer surveys. Available at http://www.globalbarometer.net/.

HIS Markit. 2014. Economics and Country Risk Index. Available at www.ihs.com/products/global-insight/index.aspx.

Holmberg, Sören, and Bo Rothstein. 2011. Correlates of Democracy. The Quality of Government Institute, University of Gothenburg, Working Paper 10.

Horton, Lynn. 2012. Is World Bank "Good Governance" Good for the Poor? Central American Experiences. *Comparative Sociology* 11: 1–28.

Hulme, David, Antonio Savoia, and Kunal Sen. 2015. Governance as a Global Development Goal? Setting, Measuring and Monitoring the Post-2015 Development Agenda. *Global Policy* 6 (2): 85–96.

International Development Law Organization. 2014. Policy Statement: Conflict Prevention, Post-Conflict Peacebuilding and Promotion of Durable Peace, Rule of Law and Governance. Eighth Session of the Open Working Group on the Sustainable Development Goals. Available at http://www.idlo.int/news/policy-statements/conflict-prevention-post-conflict-peacebuilding-and-promotion-durable-peace-rule-law-and-governance.

Kanie, Norichika, Naoya Abe, Masahiko Iguchi, Jue Yang, Ngeta Kabiri, Yuto Kitamura, Shunsuke Managi, et al. 2014. Integration and Diffusion in Sustainable Development Goals: Learning from the Past, Looking into the Future. *Sustainability* 6 (4): 1761–1775.

Kaufmann, Daniel, Aart Kraay, and Massimo Mastruzzi. 2010. *The Worldwide Governance Indicators: Methodology and Analytical Issues*. Washington, DC: World Bank.

Khan, Mushtaq H. 2007. Governance, Economic Growth and Development since the 1960s. United Nations Department of Economic and Social Affairs Working Paper 54. New York: United Nations Department of Economic and Social Affairs.

Kurtz, Marcus J., and Andrew M. Schrank. 2012. Perception and Misperception in Governance Research: Evidence from Latin America. In *Is Good Governance Good for Development?*, ed. Jomo Kwame Sundaram and Anis Chowdhury, 71–99. New York: Bloomsbury Academic.

Kwon, Huck-ju, and Eunju Kim. 2014. Poverty Reduction and Good Governance: Examining the Rationale of the Millennium Development Goals. *Development and Change* 45 (2): 353–375.

Leftwich, Adrian. 1994. Governance, the State and the Politics of Development. *Development and Change* 25 (2): 363–386.

Meon, Pierre-Guillaume, and Laurent Weill. 2010. Is Corruption an Efficient Grease? *World Development* 38: 244–259.

Millennium Challenge Corporation. 2014. Government Effectiveness Indicator. Available at http://www.mcc.gov/pages/selection/indicator/government-effectiveness-indicator.

Mo Ibrahim Foundation. 2014. Ibrahim Index of African Governance. Available at http://www.moibrahimfoundation.org/iiag-methodology/.

Mungiu-Pippidi, Alina. 2006. Corruption: Diagnosis and Treatment. *Journal of Democracy* 17: 86–100.

Nilsson, Måns, and Åsa Persson. 2012. Can Earth System Interactions Be Governed? Governance Functions for Linking Climate Change Mitigation With Land Use, Freshwater and Biodiversity Protection. *Ecological Economics* 75: 61–71.

OECD, Organisation for Economic Co-operation and Development. 2014a. DAC Peer Review. Available at http://www.oecd.org/dac/peer-reviews/.

OECD, Organisation for Economic Co-operation and Development. 2014b. Social Institutions and Gender Index. Available at http://genderindex.org/.

Open Working Group. 2014. Open Working Group on Sustainable Development Goals. Letter from the Co-Chairs, February 21. Available at http://sustainabledevelopment.un.org/focussdgs.html.

Overseas Development Institute. 2013. Are We Making Progress with Building Governance into the Post-2015 Framework? Available at http://www.odi.org.uk/publications/7295-progress-governance-post-2015-millennium-development-goals-mdgs. Accessed April 1, 2014.

Overseas Development Institute. 2014. Governance Targets and Indicators for Post 2015: An Initial Assessment. Available at http://www.odi.org.uk/sites/odi.org.uk/files/odi-assets/publications-opinion-files/8789.pdf. Accessed April 1, 2014.

Right to Information Europe and Center for Law and Democracy. 2014. Global Right to Information Rating. Available at http://www.rti-rating.org/.

Rosenau, James. 1995. Governance in the Twenty-first Century. *Global Governance* 1 (1): 13–43.

Sachs, Jeffrey D., and John W. McArthur. 2005. The Millennium Project: A Plan for Meeting the Millennium Development Goals. *Lancet* 365 (9456): 347–353.

Stiglitz, Joseph E. 2008. Is There a Post-Washington Consensus Consensus? In *The Washington Consensus Reconsidered: Towards a New Global Governance*, ed. Narcís Serra and Joseph E. Stiglitz, 41–56. New York: Oxford University Press.

Sundaram, Jomo Kwame, and Anis Chowdhury. 2012. Introduction: Governance and Development. In *Is Good Governance Good for Development*, ed. Jomo Kwame Sundaram and Anis Chowdhury, 1–28. New York: Bloomsbury Academic.

Sustainable Development Solutions Network. 2013. An Action Agenda for Sustainable Development: Report for the UN Secretary-General. Available at http://unsdsn.org/resources/publications/an-action-agenda-for-sustainable-development/ Accessed April 1, 2014.

Sustainable Development Solutions Network. 2015. Leaving No One Behind: Disaggregating Indicators for the SDGs. Available at http://unsdsn.org/wp-content/uploads/2015/10/151026-Leaving-No-One-Behind-Disaggregation-Briefing-for-IAEG-SDG.pdf. Accessed May 1, 2016.

Teorell, Jan, Carl Dahlström, and Stefan Dahlberg. 2011. The QoG Expert Survey Dataset. The Quality of Government Institute, University of Gothenburg. Available at: http://qog.pol.gu.se.

Thomas, M. A. 2010. What Do the Worldwide Governance Indicators Measure? *European Journal of Development Research* 22: 31–54.

Transparency and Accountability Initiative. 2014. Governance and the Post-2015 Development Framework: A Civil Society Proposal. Available at http://www.globalintegrity.org/wp-content/uploads/2014/01/CSO-position-on-Post-2015-and-governance-Jan-2014-hi-res-version1.pdf. Accessed April 1, 2014.

Transparency International. 2014. Corruption Perception Index.

Ugur, Mehmet. 2014. Corruption's Direct Effects on Per-Capita Income Growth: A Meta-Analysis. *Journal of Economic Surveys* 28 (3): 472–490.

UNDESA, United Nations Department of Economic and Social Affairs. 2014. Technical Support Team Issues Brief: Conflict Prevention, Post-conflict Peacebuilding and the Promotion of Durable Peace, Rule of Law and Governance. United Nations Department of Economic and Social Affairs. Available at http://sustainabledevelopment.un.org/content/documents/2639Issues%20Brief%20on%20Peace%20etc_FINAL_21_Nov.pdf.

UNDP, United Nations Development Programme. 2014. Governance for Sustainable Development. Integrating Governance in the Post-2015 Development Framework. Discussion paper. Available at http://www.worldwewant2015.org/node/429902. Accessed April 1, 2014.

UNEP, United Nations Environment Programme. 2012. 21 Issues for the 21st Century: Result of the UNEP Foresight Process on Emerging Environmental Issues. Nairobi: United Nations Environment Programme.

UNGA, United Nations General Assembly. 2015. Transforming Our World: The 2030 Agenda for Sustainable Development. Draft resolution referred to the United Nations summit for the adoption of the post-2015 development agenda by the General Assembly at its sixty-ninth session. UN Doc. A/70/L.1.

United Nations. 2013. A New Global Partnership: Eradicate Poverty and Transform Economies Through Sustainable Development. The Report of the High-level Panel of Eminent Persons on the Post-2015 Development Agenda. Available at http://www.un.org/sg/management/pdf/HLP_P2015_Report.pdf. Accessed April 1, 2014.

United Nations. 2014. UN Rule of Law Indicators. Available at http://www.un.org/en/events/peacekeepersday/2011/publications/un_rule_of_law_indicators.pdf.

Van Doeveren, Veerle. 2011. Rethinking Good Governance: Identifying Common Principles. *Policy Inquiry* 13: 301–318.

Wilkin, Sam. 2011. Can Bad Governance be Good for Development? *Survival* 53: 61–76.

Woods, Ngaire. 2000. The Challenge of Good Governance for the IMF and the World Bank Themselves. *World Development* 28: 823–841.

World Bank. 1989. *From Crisis to Sustainable Growth. Sub-Saharan Africa: A Long-term Perspective Study*. Washington, DC: The World Bank.

World Bank. 2014. World Governance Indicators. Available at http://info.worldbank.org/governance/wgi/.

World Justice Project. 2014. Rule of Law Index. Available at http://worldjusticeproject.org/rule-of-law-index.

World Values Survey Association. 2014. World Values Survey. Available at http://www.worldvaluessurvey.org/.

5 Measuring Progress in Achieving the Sustainable Development Goals

László Pintér, Marcel Kok, and Dora Almassy

The momentum offered by the Sustainable Development Goals could help find a common voice for diffuse measurement efforts. Measurement has become one of the important elements of the means of implementation, with both the UN Statistical Commission and Division playing key roles. However, this is not the first time the United Nations and governments have embarked on the construction of sustainable development indicator systems: Three editions of detailed methodological guidance were prepared and made available to member states earlier under the UN Commission on Sustainable Development (UN 2007), without achieving a convincing breakthrough in the development or mainstreaming of these indicators in decision making. A breakthrough was missed, not simply because of technical problems with the indicators put forward by these efforts or the ever-present problems with data availability. The work was limited to the technical and statistical level, and without sufficient political support and public interest, stopped short of addressing the various governance-related aspects of how the new metrics would change policies, policy implementation, and accountability regimes. Technical and statistical issues, of course, continue to persist and need addressing, but a breakthrough cannot be expected unless the work on measurement is also seen through the broader lens of governance and political economy.

Observation, measurement, and assessment are integral parts of strategic management and governance and essential for recognizing, understanding, and addressing sustainability-related problems. From individuals and society at large to businesses, governments, civil society, and multilateral organizations, measurement matters because it contributes information to decision making, whether related to understanding past problems, managing in the present, or preparing for and exploring the state of the planet.

The importance of measurement, and specifically the role of indicators in sustainable development, highlighted in chapter 40 of Agenda 21, has been recognized and has grown into a global movement since the late 1980s. They were also highlighted in the report of the 2012 UN Conference on Sustainable Development, where indicators have been recognized as a crosscutting issue and a key element of follow-up. Attention to measurement in the broader sustainability context is no longer restricted to public monitoring and reporting, but also includes the private sector through corporate reporting and civil society organizations tracking and publishing indicators and data collected by distributed networks of individual citizen observers (Herzig and Schaltegger 2011; Conrad and Hilchey 2011). This chapter argues that bringing the sustainability context to bear on measurement involved not only recognizing measurement's role in agenda setting, implementation, and reflexive evaluation, but it also placed the *problematique* of measurement in a policy and political arena that went beyond its inherently technical character.

Measurement can indeed be thought of as a technical exercise focused on gathering and presenting evidence using a wide range of monitoring, statistical survey, and remote sensing methods. However, associating measurement with the broader context of sustainability raised questions not only about its methods and instruments, but about its underlying subject—*what* is being measured, *why,* and by *whom*. Beyond the technical level, measurement in the context of the sustainability discourse became an entry point to questioning norms, values, and power structures that underpin ideas of what is being measured. This implies that indicators not only assess the status and progress of the world based on values important for human well-being, but they can also serve as a basis for society's conceptualizing and reconceptualizing sustainability and well-being (Meadows 1998).

Many of the policy arenas where sustainability, and by extension the Sustainable Development Goals, play a role require constructing transformation and transition agendas, necessary in light of the growing risks associated with business-as-usual pathways (Loorbach and Rotmans 2006; van Vuuren and Kok 2012; Özkaynak et al. 2012). Targets and indicators based on politically legitimized goals could put the construction of transition pathways on a more tangible footing. The work could build on targets and indicators already in use, as demonstrated by several illustrative efforts to develop indicators for Sustainable Development Goals (for example, Pintér, Almássy, and Hatakeyama 2014; Sustainable Development Solutions Network 2015). At the same time, relying only on the measurement systems and institutions that underpinned the development patterns leading to today's clearly unsustainable trends, as reported by most global sustainability-related assessments, is not tenable.

The need for stronger implementation and accountability in internationally agreed goals coincides with calls for a revolution in the generation and use of socioeconomic and environmental data in the reorientation of the purpose and practice of development. The Stiglitz-Sen-Fitoussi Commission ("Stiglitz Commission") on Measuring Economic Performance and Social Progress (Stiglitz, Sen, and Fitoussi 2008), the UN Secretary-General's Independent Expert Advisory Group on a Data Revolution for Sustainable Development (UN 2014b), and "beyond GDP" (European Commission 2007) initiatives are among the most prominent recent examples of efforts that try to articulate and shape a broader measurement-reform agenda.

This chapter hypothesizes that there are synergies and a convergence of interests between global measurement reform and the setting and implementation of global goals, represented at the highest level by the Sustainable Development Goals. We further argue and support earlier suggestions that understanding and capitalizing on these synergies requires considering sustainability-related measurement and the uses of measurement instruments as key aspects of governance (Hezri and Dovers 2006; Boulanger 2007). Going beyond statistical, communication, and management dimensions, we propose using a broader political economy lens to understand how measurement interacts with the interests and choices of key actors and institutions as they weigh the implications of constructing implementation agendas around the Sustainable Development Goals. When setting up measurement systems, it will be important to ensure the indicators chosen are relevant for different actor groups (such as business, issue-area networks, cities, supply chain initiatives, citizen groups, and others) that are willing to share the responsibility for implementing the Sustainable Development Goals. Agreeing on a set of universally relevant targets and related indicators and putting in place measurement systems as part of subsequent implementation at subglobal levels is where the general intentions of sustainability are demystified and the general intentions of Sustainable Development Goals become concrete, so that actors responsible for delivering solutions can act on them and be held accountable (Pintér 1997; Hajer et al. 2015).

Therein lies a challenge and an opportunity: How can the measurement-reform agenda that is as broad and universal as global goals can be rearticulated to live up to its transformative potential in the sustainability transition? For effective implementation of Sustainable Development Goals, alignment and convergence with other measurement and indicator processes—at the national, ecosystem, business, supply-chain, and product levels—will be necessary to find traction and contribute to making sustainability the new normalcy.

The chapter will discuss a number of ways in which measurement choices intersect with the interests and choices of key actors and institutions, with implications for the prospects of making progress on the Sustainable Development Goals. Sound implementation choices will depend on ensuring that the indicators are relevant and understood by different actor groups, and that they help those groups take responsibility for implementing the Sustainable Development Goals because they highlight their particular role in contributing or not contributing to progress. Therefore, monitoring of Sustainable Development Goals must become an integral and operational part of the 2030 Agenda for Sustainable Development.

If an integrated conceptual framework for Sustainable Development Goals and underlying targets that pays attention to interlinkages between different goals is developed, it would also need to be explicitly reflected in monitoring the goals and linked to national and subnational monitoring and accounting frameworks.

This would also provide an opportunity to bring economic monitoring, accounting, and reporting frameworks in line with sustainable development reporting, and thus an opportunity to redefine national development metrics, which continue to put too much weight on measuring economic growth, and place a larger emphasis on environmental sustainability and social metrics and dimensions.

A Governance Lens on Transforming Measurement Systems

Measurement is conceptualized here as a key aspect of governance to help people qualitatively and quantitatively articulate, chart, and navigate the development of complex socio-ecological systems over time and on all levels. While in the sustainable development discourse the most common aspect of measurement is related to the question of what indicators to use, a more sophisticated analysis must also recognize other dimensions whose role and function need to be considered if measurement is to contribute according to its potential in advancing sustainability (Pintér et al. 2012).

Measurement systems tend to be integral and deeply embedded, and are thus also conservative, slow-changing aspects of governance. This is most clearly demonstrated in the case of official statistics. In order to help long-term analysis, statistical monitoring, measurement, and reporting have maintained a significant level of consistency over time and space since the Second World War. With the growth of the global economy, there has been significant interest in the measurement of globalization, in particular how the system of national accounts can adapt to the new economic realities (for example, Smith 2003; Lynch and Clayton 2003). While it is not commonly thought of as such, however, the establishment of measurement systems and metrics such as gross domestic product (GDP) and gross national product were not passive instruments in the spread of globalization, but actively contributed to today's general understanding of national wealth and development. In the macroeconomic sphere, the measurement of debt, deficit, balance of payment, export-import figures, and derived metrics such as the credit rating of national economies were and still are key determinants of countries' access to capital markets, technology, and other aspects of production and consumption. They helped create a normalcy with standardization and comparability essential for determining the basic economic status of a country and assessing creditworthiness or aid eligibility. Crucially, this also contributed to spreading and cementing value systems and mindsets that underpinned the unparalleled growth of the material economy over the last several decades. Beyond their strictly economic impacts, other statistical measures play a role in promoting shifts toward more people-focused development. With the introduction of the United Nations Development Programme (UNDP) Human Development Index in 1990, an international measure was launched to compare countries on their health and education performance and per capita income and to encourage governments to consider measures beyond the GDP. Today, many countries integrate this measure into their national monitoring system, assessing their progress and performance compared to other countries.

Parallel to the development of common measurements in the economic and social spheres, in the environmental dimension the development of scientifically credible measurement, for example of the depletion of the ozone layer or the greenhouse-gas balance of the atmosphere, was essential for diagnosing problems. While not as prominent as economic indicators, in their own spheres environmental metrics became integral to the functioning of various multilateral agreements, as is the case of data on greenhouse gas emissions related to the measurement, verification, and reporting regimes of climate change–related agreements as well as the System of Environmental-Economic Accounting as part of national statistics and natural capital accounting (Breidenich and Bodansky 2009; Herold and Skutsch 2009).

Current measurement systems can, however, continue to serve as a force in governance to maintain—for the time being—the illusion of stability in the status quo, essentially by leaving unaccounted for social and natural capital deficits and debt resulting from the dominant mode of development. This is a risky proposition, considering scientific evidence that the current condition and trajectory of the earth system has entered uncharted territory with no analogue in the history of the planet (Pronk 2002). Recent analyses have shown that full costs of inaction regarding environmental problems are exceedingly high (Stern 2007; TEEB 2010) and that the earth system is already functioning in an unsafe operating space due to the breaching of some key planetary boundaries and falling behind on ensuring a minimum social floor for a significant part of humanity (Rockström et al. 2009; Raworth 2012). As an alternative, the no-analogue state of the earth system calls for no-analogue response mechanisms, including novel approaches to governance. Measurement and its associated monitoring, reporting, and verification functions are of particular interest because of their high-leverage, transformative potential, but also because of the potential for synergies that over the past decades built up among measurement-reform initiatives at multiple scales and within different actor groups (governments, corporations, nongovernmental organizations, citizens science, and so forth). The promise of the Sustainable Development Goals from the political perspective is that they can help provide a strongly mandated platform for diverse measurement and reporting efforts to speak with a common voice, to have more credible accountability mechanisms, and to shift the focus of governance toward the restructuring needed to implement the goals and to achieve a broader transformation—what implementing the goals would inherently involve.

In order to capitalize on the need for reporting on Sustainable Development Goals, and to also advance the broader agenda of global measurement reform, we argue that beside the usual technical aspects, the case also has to be constructed for the governance dimensions of measurement reform. We further argue that this means the need for the simultaneous consideration of the conceptual approach, key actors, and their interests, the mechanisms, and institutions involved, and finally the instruments of measurement. While the instruments of measurement come last in this list, it is these that are often the focus of measurement efforts, to the neglect of the other dimensions. We provide an outline of these neglected dimensions below, central for a governance approach.

Concepts

This dimension includes the worldviews and mindsets involved in defining what is important for society to measure, but also the role of measurement in decision making. The conceptual foundations for measurement often involve the development of theoretical frameworks that reflect underlying worldviews. Broad concepts such as sustainable development, well-being, and progress measurement often serve as starting points. In the case of the Sustainable Development Goals, aside from the understanding that sustainability must cover both socioeconomic and environmental aspects of development and be universally relevant, no formal framework was used as a starting point in the official process. A similar lack of a framework contributed to the fragmentary representation of the environment in the Millennium Development Goals. While the categories of Sustainable Development Goals that emerged can be considered a thematic framework, emerging from a broader societal consultation, the framework has no structure and hierarchy that would help identify priorities and interlinkages that are essential from the implementation point of view.

At the conceptual level, particularly important will be the relationship between the measurement of factors traditionally associated with "economic growth" such as per capita income, consumption, production, various metrics of macroeconomic performance, and the indicators related to the costs and risks of growth. While the conceptualization that the Sustainable Development Goals must provide a means for navigating human development in a *safe and just zone* (Raworth 2012) has gained some acceptance, there is very limited understanding of the possible trade-offs this could entail. Ultimately, conceptual inconsistencies related to Sustainable Development Goals and their implementation need to be reconciled at the level of indicators, but also translated into credible narratives on measures for moving toward Sustainable Development Goals, targets, and indicators. What the dominant narratives are, who constructs them, and how they relate to indicators can have a major influence on what gets implemented and how.

Actors

While measurement, strictly speaking, is often considered the domain of technical (for example, statistical, rating, remote sensing) agencies, the number and type of actors with stakes in the process and its results goes well beyond that. From a broader perspective of the Sustainable Development Goals, this includes actors whose interests—political, economic, or other—are affected by what is being measured and how. The clearest illustration for this is the tendency of measurement-reform initiatives that aim to create "sustainable development indicators" at various levels to involve a wide range of actors in their work and the relative ease with which involvement can be rationalized and realized; all those whose performance is assessed can recognize how the measurement approach used can affect the conclusions of such assessments. Actors and interests may also align with maintaining the status quo of measurement—enough to think, for instance, of the reliance of the financial markets on macroeconomic data that underpin the assessments of quasi-oligopolistic credit rating agencies. While these examples illustrate the tremendous institutionalized commitment to both generating and using the information, they also illustrate areas where concerns about faulty conceptualizations and methods indicate the need for change (Weber, Scholz, and Michalik 2010; White 2012; Costanza, Hart, and Posner 2009).

Mechanisms and Institutions

Changing normally conservative measurement systems requires mechanisms with a mandate agreed by relevant actors and accepted as legitimate by society. This can take different forms, from established review mechanisms of statistical agencies—such as the recent adoption of a revised framework for environmental statistics through the UN Statistical Commission and the attempt to develop more integrated social, environmental, and economic accounts—to grassroots initiatives with an ambition to develop crosscutting indicator systems at the community, national, or other levels. Starting at a larger scale in the early 1990s, the "socialization" or opening up of the field of indicator development to actors and institutions beyond statistical agencies inspired the establishment of many new initiatives with varying levels of formalization and stability, and often significant differences in approach. In contrast with more deeply established indicator systems, such as economic and trade-related components of national accounts, more recent systems such as those related to the environment may involve less formalized reviews and thus evolve with more flexibility (Smith, Pintér, and Thrift 2013). Moving toward a broader paradigm of measurement may call into question the adequacy of existing mechanisms as well as the mandate and capacity of leading institutions and their relationship with partners and audiences, as discussed by the UN Secretary-General's Independent Expert Advisory Group on a Data Revolution for Sustainable Development (UN 2014a).

Instruments

The instruments of measurement here refer mainly to the data, indicators, and related reporting mechanisms that are used to actually track progress. Identifying, changing, or adjusting them is often a key, and the most visible, objective of measurement initiatives. A broader perspective on the instruments of measurement could also include data acquisition mechanisms, without which indicators cannot be operationalized. Establishing and maintaining such capacities can be time-consuming and expensive, though the spread of information technologies and crowdsourcing through social networks is broadening the options. The same applies to reporting mechanisms where static, paper-based products are being increasingly replaced or complemented by the use of dynamically evolving, interactive tools. Through real-time data access and customized searches and presentations, these tools can increase transparency and create a more level playing field in terms of generating, analyzing, and providing access to information, with implications for vested interests and power structures. The recent UNEPLive platform, from the UN Environment Programme (UNEP), is an example of an information system that, besides improving access to data, is also intended to contribute to the synchronization of ontologies and support multiscale assessments.

Besides supply-side considerations, a governance perspective on changing measurement systems must also take into account the demand side. Uses (and misuses) of measurement can and do spring up around regularly updated and available indicators, from carbon emissions to unemployment figures, the latest GDP, or daily fluctuations of the stock market. Indicator use in society can be pervasive, and at some level affects all actors, sectors, and dimensions of environment and development. A wide range of public- and private-sector actors alike build strategies and operations around present and expected indicators, and thus create a dependency on and interest in their routine availability. Changing the metrics can call into question established patterns of decision making (Meadows 1998). This underlines the potential difficulty of changing metrics, but also their high leverage in initiating cascading effects once change does happen. The number of actors, the diversity of interests, and the high cost and potentially far-reaching implications for established routines may explain why the sustainability measurement agendas of the past often had limited or at least unclear impacts. When China embarked on an ambitious experiment to construct and introduce a "green GDP," the initiative promised to be pioneering by integrating the measurement of environmental performance into the mainstream of national accounts instead of creating a set of parallel indicators. The decision to eventually—if temporarily—abandon the project illustrates the underappreciation of the conceptual and technical complexity of the idea. But even more, it also shows the potential political complexity, as a GDP adjusted with environmental costs could have called into question the validity of economic performance figures at the national and subnational levels, affecting vested interests of both domestic and foreign actors—witness the nervousness of the markets and policymakers to news about lower than expected growth rates in China or elsewhere (Li and Lang 2010). It would have also revealed the limited understanding of how environmental costs could be integrated into the national accounts.

A broad reform of measurement systems as a component of implementation of the Sustainable Development Goals would clearly be a promising—yet as historic experience shows—complex process. The central issue is not simply adjusting the metrics of progress—it is about opening up and making more transparent the debate on what progress is, whose interests it serves, and with what results. Without considering both the supply and demand side of measurement and understanding how the change from past and present theories, practices, and tools can happen, measurement of Sustainable Development Goals would amount to little more than a technical progress-reporting exercise. To realize its transformative potential, it is useful to build on what has been learned from best-practice principles for sustainable development measurement and assessment (Pintér et al. 2012). They include the importance of a long-term perspective and visions that should underpin Sustainable Development Goals and transition pathways with intermediate goals. They call for stakeholder participation in order to build the conceptual frameworks and thematic focus of measurement systems around shared societal values and in the process build and strengthen ownership. Related to this, there is a need for preserving the broad scope that underlies the perspective of sustainable development, while recognizing systemic connections between the many actors and scales that are at play in determining higher-level and long-term outcomes. And they require a learning-adapting approach as an inherent aspect of strategy and policy (Swanson and Bhadwal 2009). Indicators for Sustainable Development Goals and their assessment would play a key role in this by providing both short- and longer-term feedback on the unfolding of multilayered, complex programs for implementation of the goals.

Past Experiences in Measuring Progress toward Global Goals

In this section we apply the four-dimensional framework outlined above to existing global goals, such as those found in multilateral environmental agreements and the earlier Millennium Development Goals. This analysis helps to understand the characteristics and the evolution of progress measurement systems. The retrospective analysis reveals that while the underlying concept of the measurement framework has remained unchanged over time, the range of the actors involved has broadened, and the interaction among institutions has become more intensive and coordinated. Yet despite the growing momentum, alternative measurement approaches have not replaced those associated with the ruling economic and sociopolitical paradigms, but have become coexistent with it.

Concepts

As for *conceptual approach*, several frameworks for measuring progress toward sustainable development issues have emerged over the last three decades. An early, extensively used framework is the pressure-state-response model, with several variants, including the activity-impact-response model developed by the UN Statistical Division, the driving force-state-response and driving force-pressure-state-impact-response models (Bartelmus 1994; Meadows 1998). A capital-based approach, developed originally by the World Bank, is based on accounting for the maintenance of a country's national wealth and considers four basic types of resources as economic, natural, human, and social capital (UNECE 2009). Building on this approach, the combination of capital-based and well-being approaches has been suggested, linking ultimate ends, intermediate ends, intermediate means, and ultimate means (from a well-being point of view) to the four different types of capital (Meadows 1998).

Other key conceptual approaches include indicator sets structured around thematic categories, such as the sustainable development indicators developed through an international process coordinated by the earlier UN Commission on Sustainable Development. The majority of the indicators for Millennium Development Goals also build on the set from the UN Commission on Sustainable Development and individual indicators that measure progress toward commitments in various international environmental and development agreements. More recently developed green economy and green growth indicators supported by UNEP and the Organisation for Economic Co-operation and Development (OECD) built on the earlier work and follow a thematic conceptualization (UN DESA 2007; OECD

2011; UNEP 2012). These models continue to coexist and are used in the context of various assessments, reports on Millennium Development Goals, and outlooks.

Actors

Traditionally, measurement of global goals followed a top-down approach, where international agencies collected, assessed, organized, and analyzed data submitted by national statistical agencies and ministries. This is, for instance, the case for the measurement of ozone-depleting substances and greenhouse gas emissions under the Vienna Protocol, the UN Framework Convention on Climate Change, and the Convention on Biological Diversity. In the Convention on Biological Diversity, the Strategic Plan for Biodiversity 2011–2020 contains a quite elaborate set of goals and targets, with a related set of indicators that also includes broadening of the measurement system to reflect ecosystem values to society (Secretariat of Convention on Biological Diversity 2010). Similarly, progress on education and health goals is tracked by international agencies, such as UNESCO and WHO. Data collection and reporting efforts have often followed a top-down, silo approach, which is not conducive to the identification and integrated assessment of interlinkages among goals in different environmental domains and various socioeconomic aspects.

The measurement of progress and reporting on Millennium Development Goals stood out at the global level as a broader effort, with a potential to bring together progress reporting on global goals to a common reporting format and common tools applicable to diverse countries. To ensure consistency and robustness of data collection for Millennium Development Goals, the process required the involvement of a wider variety of actors at the global and subglobal scales. In addition to technical agencies, nongovernmental organizations and private entities were also engaged in developing countries to which the Millennium Development Goals applied.

Mechanisms and Institutions

In contrast with other global measurement and reporting systems such as the World Development Indicators, progress reporting on Millennium Development Goals involved not only reporting the status of indicators but connecting them to goals and targets. Data for global reporting on these goals came from different sources, with the resulting inconsistencies representing a significant constraint for reporting and analysis. To coordinate the process of data collection, analysis, and reporting among the various actors, the Interagency and Expert Group on indicators for Millennium Development Goals was established, bringing together the UN Secretariat and other UN agencies; international, regional, and subregional organizations; government agencies; national statistical bodies; and development institutions.

The intention was to use measurement and reporting results to help policy makers prioritize issues in plans and strategies, assess progress or deterioration in key issues, and highlight hotspots where urgent action is needed. For example, many view the indicators for Millennium Development Goals as important for identifying, prioritizing, and integrating development goals into national planning processes. In addition, in many cases the indicators became part of annual monitoring activities and accompanied traditional economic indicators. However, due to the focus on thematic priorities of Millennium Development Goals instead of a synthetic assessment and major gaps in the issues covered (for example, in the areas of energy and the environment or the financial system), reporting on Millennium Development Goals fell short as a mechanism for overall progress reporting.

Beyond the Millennium Development Goals, as a result of the broader measurement-reform movement, social and environmental data collection, monitoring, and reporting efforts have spread and strengthened over the last two decades. In some cases, monitoring capacity also gradually improved. Global integrated assessment reports such as UNEP's Global Environment Outlook and UNDP's regular Human Development Reports provided more interpretive analysis than indicator reports. However, because of their primary focus on the environment in the case of the Global Environment Outlook and the human dimension in the case of the Human Development Reports, despite the "integrated" character, they fell short of a global sustainable development report.

Instruments

Data for stand-alone global goals and indicators on Millennium Development Goals have been collected through traditional statistical and remote-sensing agencies and methodologies. However, human and financial resource constraints, methodological differences, lack of data access, and uneven data quality persist. Due in part to advances in information technology, decentralized, citizen-driven monitoring and reporting have been spreading and producing increasing amounts of data. While the results of such efforts and the explosive availability of "big data" have not yet been systematically used in reporting on global goals, attention toward bottom-up approaches for collecting statistical indicators has seen marked growth since the beginning of the new millennium (Stiglitz, Sen, and Fitoussi 2008, Pintér, Hardi, and Bartelmus 2005). The UN Secretary-General's Independent Expert Advisory Group on a Data Revolution for Sustainable Development represents a recent high-level effort to bring political attention to the importance of data issues as a strategic priority (UN 2014a).

While the coverage, credibility, and utility of these methods varies from initiative to initiative, improving their coherence could considerably increase their usefulness and use in the future. On the output side, sustainable development indicator sets have been introduced or expanded by governments at different levels. Recognizing the importance of taking local circumstances into consideration, many regional governments and municipalities have been developing their own indicators and reporting products. In addition, many nongovernmental organizations have embarked on developing their own indicator sets, often developed for specific issue domains (for example, environment, energy, or social development) and involving the use of composite indicators (IISD 2016).

Such bottom-up approaches have immense potential to improve both data collection and monitoring for global development objectives, yet these potentials have not yet been fully exploited. Initiatives such as the Community Indicators Consortium in North America or the OECD's Wikiprogress aim to bring together such initiatives and to provide a platform for learning and information sharing. However, more effort will be needed to address the diverse nature of these initiatives and place them in a common analytical, institutional, and perhaps political framework, where they can be utilized at a greater global scale.

Measuring Progress toward the Sustainable Development Goals

At the global level, the indicators and reporting mechanism associated with the Millennium Development Goals represent one of the often-evoked precedents for measuring progress toward global goals. While the experience with indicators on these goals is indeed relevant, there are important differences when it comes to the Sustainable Development Goals. The differences are in part in the Sustainable Development Goals' broader scope and universal applicability to all countries. The Sustainable Development Goals require navigating a wider range of highly complex and interrelated issues simultaneously. This will involve a constant process of experimentation, incremental change, revision, and reordering, in which long periods of stability may be interrupted by short periods of radical change. It is almost impossible to create optimal policies in such an unpredictable setting. What will be required instead are inclusive, learning-by-doing processes, at the many relevant places, with careful monitoring and assessment of policy effects and an ability to make critical choices and improvements consistent with the trajectories leading to long-term goals. Indicators are central instruments for these tasks.

There are also other important contextual differences in a more advanced policy agenda related to measurement and its importance in various international policy mechanisms. At the broadest level and at the international level, discourses on alternatives to some of the central metrics of performance measurement such as the GDP have gained some traction, if not yet widespread adoption, over the last decade. This could clearly be witnessed at the 2012 UN Conference on Sustainable Development, where many observers expected that new metrics for measuring broad societal progress or well-being could develop a stronger consensus in the coming years (UNCSD 2012). Measurement also emerged as an important track in some international environment-development negotiations, such as measurement, reporting, and verification in the case of the UN Framework Convention on Climate Change. At a lower level, the emphasis on measuring impact by public and private entities, such as those dealing with sustainability-related standards that apply more at the organizational, supply-chain, or ecosystem level, show that measurement is an important and probably universally relevant lever with the potential to influence policy and practice. Mechanisms such as the Global Reporting Initiative resulted in an increasingly common use of indicators in corporate reports (for example, Roca and Searcy 2012), while work has also occurred on indicators at the ecosystem level (for example, Jørgensen, Xu, and Robert 2010; Bartelmus 2015), on supply chains (for example, Hassini, Surti, and Searcy 2012) or in communities (for example, Stevens 2014). While measurement initiatives like these may play a role in the sustainability performance of a given organization, the real potential for change may be in the strategic alignment of the various indicator systems around key issue priorities, as represented by the Sustainable Development Goals (Hoekstra et al. 2014; van der Esch and Steuer 2014; Measure What Matters 2015). Indicators of Sustainable Development Goals could be particularly potent instruments for policy alignment because they have to express in a succinct and concrete form the norms, interests, and aspirations of different actors, and they can act as a bridge between the normative domain and the theater of policy implementation. We argue that the implementation of Sustainable

Development Goals could be furthered by viewing measurement of Sustainable Development Goals through the four dimensions we associated with a governance lens.

Concepts

Sustainable Development Goals require big-picture thinking and an ability to recognize priorities without losing sight of essential elements of complex socio-ecological systems. The framework accepted for the goals themselves can also serve as a starting point for the identification of targets and indicators that together allow the Sustainable Development Goals to become a universally relevant agenda.

However, the Sustainable Development Goals were identified in a political process with no higher-level formal conceptual framework that would represent a consensus view of the negotiating parties on how the issues should be structured and related to each other, or of sustainable development measurement (Pintér et al. 2012). As indicators are bound to be derived from goals, the lack of a conceptual framework may carry over to sustainability metrics. Much of the attention to measurement focused on ensuring there is adequate monitoring, data collection, and reporting, building on the Secretary-General's Synthesis Report (UN 2014c) and the report of the UN Secretary-General's Independent Expert Advisory Group (UN 2014a), which provided only limited guidance with regard to the conceptual framework for measurement. In other global efforts, the work of the Sustainable Development Solutions Network focused on 10 priority challenges. The work led by the UN Statistical Division focused on identifying indicators around the goals, adopting a thematic framework similar to the Sustainable Development Solutions Network (ECOSOC 2016; SDSN 2015).

While identifying indicators in a thematic framework around the goals themselves is necessary for tracking progress toward targets, a thematic framework alone does not explicitly differentiate between the costs and benefits of development: While it has everything that is believed to be important, it does not present the indicators as a system that clearly points out the relationships between social, economic, or ecological capital.

Preliminary work on constructing or reconstructing the Sustainable Development Goals as a system of interlinked goals, targets, and indicators at the global level has started (Le Blanc 2015), and there are also examples of goal-target-indicator systems built around a conceptual framework differentiating between ultimate means and ends (for example, Pintér, Almássy, and Hatakeyama 2014). While these illustrate that Sustainable

Development Goals and indicators can be organized or can even be developed around conceptual frameworks that are explicit about causal interlinkages between drivers and results of development, such frameworks may need to be explicitly accepted in monitoring and analysis of Sustainable Development Goals to have impact on implementation.

The framework for monitoring would need to be linked with key policy priorities and policy interests, and thus strengthen accountability at the national and subnational level. Monitoring would need to be consistent with or grow out of existing national monitoring frameworks, including frameworks for economic reporting that would need to come under the umbrella of higher-level sustainable development reports. It would be particularly important to use this opportunity to redefine the ultimate purpose and metrics of national development as human well-being as opposed to simply economic growth, as already advocated by both academic and policy schools of thought (for example, Costanza, Hart, and Posner 2009). While both the technical foundations—and now, through the Sustainable Development Goals, the political momentum—exist, introducing the necessary changes still needs a coherent and conceptually sound narrative and a transition strategy.

Actors

Due to their broader focus and universality, Sustainable Development Goals are relevant for a broader range of actors in general and a range of actors *interested in measurement* in particular. Understanding the role of agency in measurement of the Sustainable Development Goals requires adopting a supply and demand perspective, with separate "instrument constituencies" related to both (Voß and Simons 2014). There are potentially new implications for both, compared to earlier experiences. The literature on measurement has traditionally been dominated by interests in the supply side, adopting an implicit "if we build it they will come" position. More recent interest in the use and impact of indicators—illustrated, for example, by the designation of the 2016 annual conference of the US-based Community Indicator Consortium as the "Impact Summit"—however, shows the need for moving beyond this position in the case of place-based measurement initiatives, and it is even more important to do so in the case of the Sustainable Development Goals.

With regard to the supply side, as the data revolution report of the Independent Expert Advisory Group and earlier reports have shown, there is a need to strengthen the monitoring and observation capacity of actors that traditionally played such roles, including statistical and remote-sensing agencies like the UN Statistics Division or the Group on Earth Observation at the international level (UN 2014a). However, due to increasing actor interest and widespread availability of enabling technologies, monitoring and observation are now carried out by a much wider range of actors, from civil society to the private sector (van der Esch and Steuer 2014). Organized integrated monitoring and sustainability indicator efforts now have an established tradition and may be developed around a wide range of reference systems, whether a community, region, country, sector, commodity, or organization (for example, Rydin 2010; Hák, Moldan, and Dahl 2012). While many of these represent broad thematic interests, often they cover only specific segments of the overall agenda of Sustainable Development Goals. Under the Sustainable Development Goals, integrating or synthesizing sustainability information from diverse sources will receive more emphasis and call for either new actors as information suppliers or for strengthening the data provision and integrator function of existing actors. A growing number of efforts such as UNEPLive are under way that ultimately aim to provide access to observation and measurement information provided by traditional and nontraditional actors at the global level. These can support monitoring of the Sustainable Development Goals, but, at the national level, specific organizational capacities and specific platforms will likely be needed, perhaps even new institutions with a mandate to cover the needs of reporting on the goals. It is unclear whether the framework of the goals will help structure these initiatives, but it clearly has the potential to do so, given its broad thematic coverage, information needs, and political support built through the process.

Similar to the supply side, the demand-side interest in monitoring Sustainable Development Goals and their indicators will incorporate actors that have traditionally been more on the periphery of the sustainability discourse. Whether this happens depends to some extent on having open and timely access to monitoring information, and it is a matter of transparency. Publishing up-to-date indicator information is a well-covered field of interest to mainstream measurement and reporting agencies, even though it may represent new challenges in a world where data are synthesized from multiple sources, some of which may have privacy concerns.

A more challenging demand-side issue, however, is that—if the Sustainable Development Goals indeed become a universally relevant agenda—the number and type of actors interested in using their measurements will increase. With increased interest will come increased potential for conflict between the use of indicators for Sustainable Development Goals and more traditional types of indicators. Indicator use itself is multidimensional, and ranges from symbolic to political or technical uses or their combination, where the type of use will fundamentally determine whether indicators end up having an influence on actual policy and outcomes (Bell and Morse 2001; Hezri and Dovers 2006; Hildén and Rosenström 2008). Indicator use is a function of ownership, which will be in direct connection with the ownership of Sustainable Development Goals overall. In this regard, Nilsson and Persson (2015) discuss the importance of a two-track approach, where national-level, context-specific targets and indicators could ensure maximum relevance to actor interests, while a higher-level international system of indicators in a common reporting framework would ensure at least partial comparability. The two tracks would not directly overlap, but they would be aligned on issues that are relevant on all levels and through the participation of key actors and stakeholders build ownership of the SDGs.

Mechanisms and Institutions

In one of the most extensively referenced studies on the subject, Meadows (1998) observes that the *process* of indicator selection is as important as indicators themselves. In the case of the Sustainable Development Goals, measurement and indicator system development can at this stage be viewed at two different levels: global and subglobal, as also discussed by Nilsson and Persson (2015). At the global level, indicator development is centralized and closely follows the goals identified in the Open Working Group process. At the subglobal level, indicator development will be decentralized and led by countries and other subnational actors with geographic, sectoral, organizational, or other more specific interests. While similar, the procedural issues and options at these two levels are not the same. The objective of the global process is to come up with an indicator set that adheres as much as possible to common statistical standards and to set a precedent for what the key measurable dimensions of the identified Sustainable Development Goals are.

As any given goal can be measured in more than one way and as the choice of the indicators can significantly affect what targets are set, how they are implemented, and what interests they represent, measurement-system development, indicator choice, and the way indicators are selected at the global level will be highly consequential. Considering that the success of the measurement system will depend in many cases on the willingness and ability of distributed networks of subglobal actors to collect and report data, participation and transparency of the global indicator selection process will be key, as this significantly affects ownership and buy-in. The process also requires coordination and leadership, assigned to the UN Statistics Division, that can represent in a balanced way political and technical perspectives and provide a forum for the expression and resolution of different positions.

Beyond indicator selection, the broader issue of measurement-system development also needs a process for data collection, monitoring, and reporting. SDG monitoring should break with the exclusive top-down data collection and monitoring approach and also consider innovative, bottom-up, actor-group centered or community-driven initiatives. This is not only important to exploit because of the potential these alternative approaches can offer, but also has the benefit of ensuring the involvement of stakeholders in monitoring and can thus strengthen salience and buy-in. Distributed monitoring and data collection could increase societal awareness and improve ownership of goals and reduce costs. The Independent Expert Advisory Group called for a global "Network of Data Innovation Networks" and major investment in statistical and geospatial data collection, as well as for making data freely available that can contribute to the awareness of progress on Sustainable Development Goals and create incentives for innovation and improving performance.

Moreover, the measurement system for Sustainable Development Goals will need to directly support progress assessment and reporting, both of which are essential for accountability and learning. As the reporting framework evolves, global-level indicators will need to increasingly support Global Sustainable Development Reports, to be published under the authority of the United Nations (UN 2015). Given the nonbinding nature of the Sustainable Development Goals and the significant stakes in terms of investment and policy, accountability has become a politically sensitive issue during the global negotiations, with some actors calling for stronger accountability measures and others preferring flexibility. In order to make it credible, reporting would need to be timely, verifiable, independent from political interference, based on rigorous methods, and broadly and openly available. These aspects could contribute to making reporting consequential, with relevant actors—whether government, business, or civil society—weighing the results more seriously and using the information to adapt and adjust behavior.

While the measurement-system and indicator development at the global level is centralized around the United Nations and affiliated agencies, their implementation at the subglobal level will require its own processes. Many of the issues that are relevant for developing a measurement system for Sustainable Development Goals and their indicators at the global level also apply at subglobal scales, with the exception that subglobal processes can use the results of the global process—global indicator sets and reporting templates—as a starting point. However, while official statistical, remote-sensing, and reporting organizations can play a central role, at the subglobal level there is a broader diversity of actors who already run their own measurement processes and systems upon which measurement of Sustainable Development Goals can build. This can create competition for attention and open up a marketplace for products and ideas to track progress and its drivers and to outline what options are available for improving performance in the future. Both globally and within countries this opens up possibilities for divergence and convergence; divergence in the sense that parallel and competing measurement and reporting processes can reach different conclusions and promote alternative implementation strategies and pathways, and convergence in the sense that the "sense making" that is involved in these processes can promote different actors realizing common interests and the need for coherent, collaborative action.

Instruments

The instrumental dimension of Sustainable Development Goals covers a number of issues that relate to indicator choice and design, but are also linked to the instruments of indicator communication, reporting, and even indicator use. As key considerations, the instruments need to directly interface with and reflect as closely as possible the aspirations expressed through goals and targets. For a number of reasons, this represents a significant challenge for the Sustainable Development Goals. While these challenges were already known from earlier measurement initiatives, in the case of the Sustainable Development Goals, the issues are more exposed and receive more scrutiny because they represent a higher level of complexity and more explicit political interests. Many of the design choices involve difficult trade-offs, and the extended global negotiations on Sustainable Development Goals provided ample opportunities for actors to recognize that the choices made can be consequential for implementation.

One of the key considerations is that the indicators selected should directly build on goals and targets as part of a "package" (Pintér et al. 2013). As the number of goals and targets in the 2014 proposal of the Open Working Group on Sustainable Development Goals stood at 17 goals and 169 targets (UN 2014c), it was inevitable that at least initially the indicator system developed would be similarly complex. Indeed, in its early 2015 working draft report, the Bureau of the UN Statistical Commission presented a list of 304 potential indicators, not even covering all 169 targets (UN Statistical Commission 2015). While there is no objective way to define an optimal number of indicators, the need for keeping the number at a "manageable" level has long been recognized by the practitioners' community (Hardi and Zdan 1997; Mitchell 1996). The challenge in the case of the Sustainable Development Goals is that because of the close coupling of the elements of the Sustainable Development Goals package and the primacy of goals and targets over indicators, keeping the number of indicators manageable without limiting the number of goals and targets would create an inconsistency. There is, however, very minimal political will at the global level to reduce the number of goals and targets, both because of the effort that was invested in reaching consensus on the goals and targets through the Open Working Group process, and because the negotiating parties didn't want to give up hard-fought goals and targets important in their own context.

While the indicator system for Sustainable Development Goals does not have a perfect historic analogue, many earlier indicator development processes involved prioritization and the identification of a core set of indicators (for example, Department for Environment, Food and Rural Affairs 2013; European Environment Agency 2005). Prioritization of indicators based on selection criteria has also been practiced in the private sector, where key performance indicators are established components of management systems, and their operational use requires keeping them focused on main management objectives (for example, Shahin and Mahbod 2007). If identifying a core set of indicators at the global level is not politically feasible, prioritization will likely take place at the subglobal level as part of national efforts to identify goals and targets that match local context. Over time, out of this iterative contextualization process a tighter set of internationally common indicators may emerge, based on priority indicators most commonly selected by countries and other actors. Alternatively, there might also be an increase in core indicator set recommendations. While not without risks (for example, how to ensure all key global issues such as those related to planetary boundaries are covered), a bottom-up pathway to common global indicators is consistent with views that emphasize the learning-by-doing, evolutionary nature of indicator system development (for example, Meadows 1998; Hák, Moldan, and Dahl 2012, Pintér, Hardi, and Bartelmus 2005).

With regard to instrumentation, another question is the development and use of aggregate or composite indices in decision making (for example, Nardo et al. 2005; Böhringer and Jochem 2007; Mayer 2008). Given the nature of Sustainable Development Goals and their targets and the need for guiding policy and measuring, present efforts focus on developing indicators rather than indices. Considering the challenges associated with the selection of more transparent indicators, developing a "superaggregate" sustainability index was not on the political agenda at the time of the negotiations. However, there are at least three reasons for considering indicator aggregates as a serious possibility at some point.

First, outside of the sustainability domain there are many mainstream indices that play a central role in decision making. While there is a significant body of literature that studies and critically reviews the use, non-use, and misuse of sustainability indicators and indices (for example, Lehtonen 2015; Böhringer and Jochem 2007), there are many examples of indicator systems and aggregate indices—such as stock market indices, GDP, or trade-related indicators—that profess no linkage to the notion of sustainability and that *are* being used, even though their distortions are known (for example, Kubiszewski et al. 2013; Castro-Martínez, Jiménez-Sáez, and Ortega-Colomer 2009). If the 2030 Agenda for Sustainable Development and the Sustainable Development Goals are to have an effect on policy making, they need to present credible and robust alternatives to measurement systems that in the past marginalized sustainability perspectives. Robustness means that the indicator system probably needs to interface with the System of Environmental-Economic Accounting, which is based on the System of National Accounts and has been formally adopted by the UN Statistical Commission. Considering that many of the presently used measurement systems include aggregates, the development and use of indices related to the Sustainable Development Goals should at some point also be considered.

Second, even if not used as extensively as mainstream economic indices, over the previous decades several aggregate indices related to some key sustainability domains have been developed, such as the Human Development Index or the Environmental Performance Index. These indices are produced by respectable organizations using standardized methods; they rely on the best available data and have considerable time series; and they are used for benchmarking and comparison, while many serve as a basis for regularly issued policy-relevant assessments. Most if not all of their component metrics have their equivalents in the proposed set of indicators for Sustainable Development Goals. With or without adjusting their structure, considering their application as part of the planning of the implementation, monitoring, and reporting system for the Sustainable Development Goals would be logical and provide continuity with measurement tools already known to decision makers. More explicitly linking them with the post-2015 context and the Sustainable Development Goals is what may be needed to give them a stronger role.

As a third consideration regarding aggregation, part of the problem with many of today's indices is that their aggregation algorithms are too complex or not sufficiently transparent, which makes them appear as black boxes to decision makers. This either limits their use (as in the case of many sustainability-related indices) or it results in misuse (as in the case of GDP) and the perpetuation of mental models in decision making that are essentially unsustainable (for example, Costanza, Hart, and Posner 2009). There are, however, possibilities for overcoming this problem by developing a conceptual architecture for indicator systems that combines singular indicators with aggregates in a transparent way. An early proposal by the government of Colombia suggested the development of a Global Dashboard for the post-2015 Development Agenda that could both serve as a presentation and decision-support tool and help track progress toward the aspirations in the Sustainable Development Goals (Government of Colombia 2013). A simultaneous presentation of sustainability indicators and aggregates has already been prepared—the Dashboard of Sustainability in the 1990s—and was later applied to the indicators for Millennium Development Goals (Joint Research Centre 2015).

Although going beyond the strictly defined field of measurement, indicators are expected to serve as an essential element of progress reporting on Sustainable Development Goals. Reporting is a key part of implementation of the goals and accountability mechanisms, and the instrumentation and methods of reporting will play a key role in their usefulness. While dashboards and interactive web-based progress-reporting platforms can play a role and provide up-to-date information, simple, indicator-based analyses alone would be inadequate to reveal deeper structural interlinkages that are essential for understanding and addressing the nuances of complex sustainability problems (Fukuda-Parr and McNeill 2015). As many of these nuances cannot even be quantified, indicators for Sustainable Development Goals and indicator-based model results would need to be complemented by qualitative information using suitable integrated assessment frameworks and reporting formats (Beisheim et al. 2015). The High-level Political Forum has been designated as the global reporting authority through the Global Sustainable Development Report (UN DESA 2013).

Parallel to global reporting, national reporting instruments and processes are required and may be the most significant level for reporting, given that implementation of the Sustainable Development Goals will be mainly centered at the country level (de la Mothe, Espey, and Schmidt-Traub 2015). Reporting at the national level may itself need to be a multilevel reporting system, with different actors leading their own thematic, geographic, sectoral, or organization-level initiatives, depending on their involvement in specific activities to implement Sustainable Development Goals. More than in the case of the Millennium Development Goals, this may need to involve other sectors than the various levels of government. Although with great differences between countries based on their levels of development, there are many cases for civil society and business-sector involvement in reporting on the Millennium Development Goals. Private-sector voluntary reporting can build on experience with corporate sustainability reports, such as the adoption of the Global Reporting Initiative standard, and even take the initiative in identifying nationally relevant Sustainable Development Goals, targets, and related indicators to scope out how business can contribute, as was the case in Hungary, where the Business Council on Sustainable Development, a national member of the World Business Council for Sustainable Development, led a national initiative on the Sustainable Development Goals (BCSDH 2015). Management standards such as ISO and the European Eco-Management and Audit Scheme also cover protocols for reporting as an integral element of management systems and could provide a logical base for tracking and reporting on indicators for Sustainable Development Goals.

Conclusion

Sustainability measurement and indicators are situated at the interface of science and policy, yet measurement is often considered a primarily scientific and technical exercise, even if indicator development is increasingly opened up to public participation, especially at the community level. As noted by McCool and Stankey (2004), even if they are all involved in the same measurement-system development process, the interests of government, business, civil society, and science vary, and they use indicators differently and for different purposes. Understanding the reasons for these differences with regard to different aspects of measurement-system development and use, and taking them into account, is important for building a strong foundation for the implementation of Sustainable Development Goals at both global and national levels.

Some ambiguities about where responsibility for the development of indicators for Sustainable Development Goals should lie have already emerged in the global process. Formally, the responsibility to develop the indicators was delegated to a technical body, the UN Statistical Commission. However, as the negotiations wore on, at the April 2015 session of the Open Working Group, countries were questioning whether indicator selection would also require stronger intergovernmental involvement. While some were ready to negotiate indicators in the same detail as the goals and targets themselves, with Bangladesh emphasizing that "all data can be manipulated without a careful framework," the fact that the task was already given to the UN Statistical Commission and the potentially large number of indicators and accompanying technical detail made detailed negotiation impractical (Lebada, Offerdahl, and Paul 2015).

This episode illustrates a number of points made in the chapter with regard to measurement systems for Sustainable Development Goals. While earlier literature (for example, Meadows 1998) is clear that indicators represent high-leverage and influential elements of complex systems, policymakers often continue to treat measurement as a purely technical exercise. The adequacy of delegating indicator development to the UN Statistical Commission was initially seen as logical, and concerns about the sufficiency of that process were raised only when delegates, already deeply entrenched in negotiations on Sustainable Development Goals, realized the possible significant implications that indicator system development entails. No similar concerns were raised during the earlier development of the UN Commission on Sustainable Development's indicator system: The process started as a technical and statistical exercise and ended as that, with limited impact on policy, and with no implementation even of the System of Environmental-Economic Accounting, despite the recommendations of Agenda 21 (UNDPCSD 1996; Pintér, Hardi, and Bartelmus 2005). Indicators for Sustainable Development Goals, on the other hand, followed the goal- and target-setting process with a clear expectation that goals would be consequential and implemented. Indicators under such conditions are no longer marginal and purely technical: They are central and political. It is for this reason that we have argued they ought to be considered under a governance framework, not simply evaluated on the match of indicators to goals or targets. We proposed in this chapter a political economy framework with concepts, actors, mechanisms and institutions, and instruments as the four main elements. We demonstrated the use of the framework for analyzing an approach to measurement retrospectively, related to the Millennium Development Goals, and showed its relevance to analyze the Sustainable Development Goals and that it can serve not only academic, but also strategic, policy purposes.

The policy relevance and political aspects of indicators are becoming even clearer as implementation of Sustainable Development Goals proceeds. Mainstream indicators such as stock market indices, economic growth figures, and employment and inflation data are powerful and influence institutional and personal behavior in fundamental ways. If indicators for Sustainable Development Goals are to make a difference, they need to compete for attention and gain relevance in the same theater of decision making as widely used economic indicators. This requires more nuanced understanding of issues not only related to technical development, but to policy uses of indicators, which until recently has been a relatively neglected area of interest of public policy scholars and indicator practitioners. Research focused on the uses of indicators reveals significant differences depending on the type of use, and shows that intended uses could already inform indicator system development (Hezri 2006; Lehtonen 2015). This will be particularly important at the country level, where most of the implementation of Sustainable Development Goals is expected.

The development of a measurement system for Sustainable Development Goals is both a novel science-policy challenge and a field that can build on the lessons from the experience with existing sustainable development indicators and from the experience of the Millennium Development Goals. It seems that developing a sustainability indicator system on the foundation of high-level goals and targets raises the stakes and may invite closer political scrutiny of both the process and results. On the other hand, the development of indicators for Sustainable Development Goals could more strategically build on two decades of experience related to the conceptualization, process design, and use of indicators.

References

Bartelmus, Peter. 1994. Towards a Framework for Indicators of Sustainable Development. United Nations Department for Economic and Social Information and Policy Analysis, Working Paper Series 7. New York: United Nations.

Bartelmus, Peter. 2015. Do We Need Ecosystem Accounts? *Ecological Economics* 118: 292–298.

Beisheim, Marianne, Hedda Løkken, Nils aus dem Moore, László Pintér, and Wilfried Rickels. 2015. *Measuring Sustainable Development: How Can Science Contribute to Realizing the SDGs? Working paper.* Berlin: German Institute for Security and International Affairs. Available at: http://www.swp-berlin.org/fileadmin/contents/products/arbeitspapiere/Beisheim-et-al_WorkingPaper_MeasuringSD.pdf.

Bell, Simon, and Stephen Morse. 2001. Breaking through the Glass Ceiling: Who Really Cares About Sustainability Indicators? *Local Environment* 6 (3): 291–309.

Böhringer, Christoph, and Patrick E. P. Jochem. 2007. Measuring the Immeasurable: A Survey of Sustainability Indices. *Ecological Economics* 63 (1): 1–8.

Boulanger, Paul-Marie. 2007. Political Uses of Social Indicators: Overview and Application to Sustainable Development Indicators. *International Journal of Sustainable Development* 10 (1): 14–32.

Breidenich, Clare, and Daniel Bodansky. 2009. *Measurement, Reporting and Verification in a Post-2012 Climate Agreement.* Washington, DC: Pew Center on Global Climate Change.

Business Council for Sustainable Development in Hungary. 2015. Action 2020. Budapest: Business Council for Sustainable Development in Hungary. Available at: http://action2020.hu/en/action-2020-hungary/.

Castro-Martínez, Elena, Fernando Jiménez-Sáez, and Francisco Javier Ortega-Colomer. 2009. Science and Technology Policies: A Tale of Political Use, Misuse and Abuse of Traditional R&D Indicators. *Scientometrics* 80 (3): 827–844.

Conrad, Cathy C., and Krista G. Hilchey. 2011. A Review of Citizen Science and Community-based Environmental Monitoring: Issues and Opportunities. *Environmental Monitoring and Assessment* 176 (1–4): 273–291.

Costanza, Robert, Maureen Hart and Stephen Posner. 2009. Beyond GDP: The Need for New Measures of Progress. The Pardee Papers, 4, The Pardee Center for the Study of the Longer-Range Future, Boston University.

de la Mothe, Eve, Jessica Espey, and Guido Schmidt-Traub. 2015. Measuring Progress on the SDGs: Multi-level Reporting. Global Sustainable Development Report 2015 Brief. New York: Sustainable Development Solutions Network. Available at: https://sustainabledevelopment.un.org/content/documents/6464102-Measuring%20Progress%20on%20the%20SDGs%20%20%20Multi-level%20Reporting.pdf.

Department for Environment, Food and Rural Affairs. 2013. Sustainable Development Indicators. London: Department for Environment, Food and Rural Affairs. Available at: https://www.gov.uk/government/uploads/system/uploads/attachment_data/file/223992/0_SDIs_final__2_.pdf.

ECOSOC, United Nations Economic and Social Council. 2016. Report of the Inter-Agency and Expert Group on Sustainable Development Goal Indicators. E/

CN.3/2016/2/Rev.1*. Available at: http://unstats.un.org/unsd/statcom/47th-session/documents/2016-2-SDGs-Rev1-E.pdf.

European Commission. 2007. Summary notes from the Beyond GDP conference. Highlights from the presentations and the discussion. Available at: http://ec.europa.eu/environment/beyond_gdp/download/bgdp-summary-notes.pdf.

European Environment Agency. 2005. EEA Core Set of Indicators—Guide. Copenhagen: European Environment Agency.

Fukuda-Parr, Sakiko, and Desmond McNeill. 2015. Post 2015: A New Era for Accountability? Presented at The Development of an Indicator Framework for the Post-2015 Development Agenda: Towards a Nationally Owned Monitoring System for the SDGs. Workshop of the United Nations Statistical Commission, February 27, 2015, New York. Available at: http://unstats.un.org/unsd/statcom/statcom_2015/seminars/post-2015/docs/Panel%201.3_FukudaParr%20.pdf.

Government of Colombia. 2013. A Global Dashboard for the New Post-2015 Development Agenda. Available at: https://sustainabledevelopment.un.org/content/documents/3621colombia.pdf.

Hajer, Maarten, Måns Nilsson, Kate Raworth, Peter Bakker, Frans Berkhout, Yvo de Boer, Johan Rockström, et al. 2015. Beyond Cockpit-ism: Four Insights to Enhance the Transformative Potential of the Sustainable Development Goals. *Sustainability* 7 (2): 1651–1660.

Hák, Tomás, Bedrich Moldan, and Arthur Lyon Dahl, eds. 2012. *Sustainability Indicators: A Scientific Assessment*. Vol. 67. Washington: Island Press.

Hardi, Péter, and Terrence Zdan. 1997. *Assessing Sustainable Development: Principles in Practice*. Winnipeg: International Institute for Sustainable Development.

Hassini, Elkafi, Chirag Surti, and Cory Searcy. 2012. A literature review and a case study of sustainable supply chains with a focus on metrics. *International Journal of Production Economics* 140: 69–82.

Herold, Martin and Margaret M. Skutsch. 2009. Measurement, Reporting and Verification for REDD+: Objectives, capacities and institutions. In *Realising REDD*, ed. Arild Angelsen, Markku Kanninen, Erin Sills, William D. Sunderlin, and Sheila Wertz-Kanounnikoff, 85–100. Bogor: Center for International Forestry Research.

Herzig, Christian, and Stefan Schaltegger. 2011. Corporate sustainability reporting. In *Sustainability Communication: Interdisciplinary Perspectives and Theoretical Foundations*, ed. Jasmin Godemann and Gerd Michelsen, 151–159. Dordrecht: Springer.

Hezri, Adnan A. 2006. Connecting Sustainability Indicators to Policy Systems. Doctoral dissertation, Canberra, Australian National University.

Hezri, Adnan A., and Stephen R. Dovers. 2006. Sustainability Indicators, Policy and Governance: Issues for Ecological Economics. *Ecological Economics* 60 (1): 86–99.

Hildén, Mikael, and Ulla Rosenström. 2008. The Use of Indicators for Sustainable Development. *Sustainable Development* 16 (4): 237–240.

Hoekstra, Rutger, Jan Pieter Smits, Koen Boone, Walter van Everdingen, Fungayi Mawire, Bastian Buck, Anne Beutling, et al. 2014. Reporting on Sustainable Development at National, Company and Product Levels: The Potential for Alignment of Measurement Systems in a Post-2015 World. Statistics Netherlands, Global Reporting Initiative, and The Sustainability Consortium.

IISD, International Institute for Sustainable Development. 2016. Compendium—A Global Directory to Indicator Initiatives. Winnipeg, MB: International Institute for Sustainable Development. Available at: https://www.iisd.org/measure/compendium/.

Joint Research Centre. 2015. Millennium Development Goals Dashboard. Available at: http://www.webalice.it/jj2006/MdgDashboard.htm.

Jørgensen, Sven E., Liu Xu, and Costanza Robert, eds. 2010. *Handbook of Ecological Indicators for Assessment of Ecosystem Health*. Boca Raton, FL: CRC press.

Kubiszewski, Ida, Robert Costanza, Carol Franco, Philip Lawn, John Talberth, Tim Jackson, and Camille Aylmer. 2013. Beyond GDP: Measuring and Achieving Global Genuine Progress. *Ecological Economics* 93: 57–68.

Lebada, Ana Maria, Kate Offerdahl, and Delia Paul. 2015. Summary of the third session of intergovernmental negotiations on the post-2015 development agenda: 23–27 March, 2015. *Earth Negotiations Bulletin* 32 (16). Available at: http://www.iisd.ca/download/pdf/enb3216e.pdf.

Le Blanc, David. 2015. Towards Integration at Last? The Sustainable Development Goals as a Network of Targets. DESA Working Paper, 141. United Nations Department of Economic and Social Affairs. Available at: http://www.un.org/esa/desa/papers/2015/wp141_2015.pdf.

Lehtonen, Markku. 2015. Indicators: Tools for informing, monitoring or controlling? In *The Tools of Policy Formulation. Actors, Capacities, Venues and Effects*, ed. A. J. Jordan and J. R. Turnpenny, 76–99. Cheltenham: Edward Elgar.

Li, Vic, and Graeme Lang. 2010. China's "Green GDP" experiment and the struggle for ecological modernisation. *Journal of Contemporary Asia* 40 (1): 44–62.

Loorbach, Derk, and Jan Rotmans. 2006. Managing Transitions for Sustainable Development. In *Understanding Industrial Transformation. Views from Different Disciplines*, ed. Xander Olshoorn and Anna J. Wieczorek, 187–206. Dordrecht: Springer.

Lynch, Robin, and Tony Clayton. 2003. Globalisation: New Needs for Statistical Measurement. *Statistical Journal of the United Nations Economic Commission for Europe* 20 (2): 121–134.

Mayer, Audrey L. 2008. Strengths and Weaknesses of Common Sustainability Indices for Multidimensional Systems. *Environment International* 34 (2): 277–291.

McCool, Stephen F., and George H. Stankey. 2004. Indicators of Sustainability: Challenges and Opportunities at the Interface of Science and Policy. *Environmental Management* 33 (3): 294–305.

Meadows, Donella H. 1998. *Indicators and Information Systems for Sustainable Development. A Report to the Balaton Group*. Hartland Four Corners, VT: The Sustainability Institute.

Measure What Matters. 2015. Monitoring for Sustainable Development: The Need for Alignment. Background Paper 3. Available at: http://measurewhatmatters.info/wp-content/uploads/2015/07/MWM-IRF-Retreat-7.-Monitoring-for-Sustainable-Development-The-Need-for-Alignment.pdf.

Mitchell, Gordon. 1996. Problems and Fundamentals of Sustainable Development Indicators. *Sustainable Development* 4 (1): 1–11.

Nardo, Michela, Michaela Saisana, Andrea Saltelli, Stefano Tarantola, Anders Hoffmann, and Enrico Giovannini. 2005. Handbook on Constructing Composite Indicators. Organisation for Economic Co-operation and Development Statistics Working Paper. TSD/DOC 3. Paris: Organisation for Economic Co-operation and Development. Available at: http://www.oecd.org/std/42495745.pdf.

Nilsson, Måns, and Åsa Persson. 2015. How Do We Get Real National Ownership of the post-2015 Agenda? Draft research note for Workshop on Implementing the Sustainable Development Goals, Arizona State University, April 25–26, 2015. Available at: http://conferences.asucollegeoflaw.com/sdg2015/files/2012/08/17-Nilsson-and-Persson-input-to-ASU-SDG-workshop-April-2015.pdf.

OECD, Organisation for Economic Co-operation and Development. 2011. *Towards Green Growth: Monitoring Progress*. Paris: OECD.

Özkaynak, Begüm, László Pintér, Detlef P. van Vuuren, Livia Bizikova, Villy Christensen, Martina Floerke, Marcel Kok, et al. 2012. Scenarios and Sustainability Transformation. In *Global Environment Outlook 5*. Nairobi: United Nations Environment Programme.

Pintér, László. 1997. De-mystifying Sustainable Development through Performance Measurement. In *Sustainable Development—Implications for World Peace*, ed. A. R. Magalhães, 61–73. Austin, TX: Lyndon B. Johnson School of Public Affairs and The University of Texas at Austin.

Pintér, László, Dóra Almássy, Ella Antonio, Sumiko Hatakeyama, Ingeborg Niestroy, Simon Olsen, and Grazyna Pulawska. 2013. *Sustainable Development Goals for a Small Planet: Connecting the Global to the National Level in 14 countries of Asia-Pacific and Europe. Part I: Methodology and Goals Framework*. Singapore: Asia-Europe Foundation.

Pintér, László, Dóra Almássy, and Sumiko Hatakeyama. 2014. *Sustainable Development Goals for a Small Planet: Connecting the Global to the National Level in 14 Countries of Asia-Pacific and Europe. Part II: Measuring Sustainability*. Singapore: Asia-Europe Foundation. Available at: http://www.asef.org/images/stories/publications/documents/ENVforum-Part_II-Measuring_Sustainability.pdf.

Pintér, László, Péter Hardi, and Peter Bartelmus. 2005. Indicators of Sustainable Development: Proposals for a Way Forward. Discussion paper. UN Division for Sustainable Development, EGM/ISD/2005/CRP.2. Available at: http://www.un.org/esa/sustdev/natlinfo/indicators/egmIndicators/crp2.pdf.

Pintér, László, Péter Hardi, André Martinuzzi, and Jon Hall. 2012. Bellagio STAMP: Principles for Sustainability Assessment and Measurement. *Ecological Indicators* 17: 20–28.

Pronk, Jan. 2002. The Amsterdam Declaration on Global Change. In *Challenges of a Changing Earth* (proceedings of the Global Change Open Science Conference, Amsterdam, The Netherlands, 10–13 July 2001), eds. Will Steffen, Jill Jäger, David J. Carson and Clare Bradshaw, 207–208. Berlin: Springer.

Raworth, Kate. 2012. A Safe and Just Space for Humanity: Can We Live Within the Doughnut? *Oxfam Policy and Practice: Climate Change and Resilience* 8 (1): 1–26.

Roca, Laurence Clément, and Cory Searcy. 2012. An Analysis of Indicators Disclosed in Corporate Sustainability Reports. *Journal of Cleaner Production* 20 (1): 103–118.

Rockström, Johan, Will Steffen, Kevin Noone, Åsa Persson, F. Stuart Chapin III, Eric Lambin, Timothy M. Lenton, et al. 2009. Planetary Boundaries: Exploring the Safe Operating Space for Humanity. *Ecology and Society* 14 (2): 32.

Rydin, Yvonne. 2010. *Governing for Sustainable Urban Development*. London: Routledge.

SDSN, Sustainable Development Solutions Network. 2015. Indicators and a Monitoring Framework for the Sustainable Development Goals. Launching a Data Revolution for the SDGs. Sustainable Development Solutions Network. Available at: http://unsdsn.org/wp-content/uploads/2015/05/FINAL-SDSN-Indicator-Report-WEB.pdf.

Secretariat of Convention on Biological Diversity. 2010. The Strategic Plan for Biodiversity 2011–2020 and the Aichi Biodiversity Targets. Decision adopted by the Conference of the Parties to the Convention on Biological Diversity at its Tenth Meeting. Available at: https://www.cbd.int/doc/decisions/cop-10/cop-10-dec-02-en.pdf.

Shahin, Arash, and M. Ali Mahbod. 2007. Prioritization of Key Performance Indicators: An Integration of Analytical Hierarchy Process and Goal Setting. *International Journal of Productivity and Performance Management* 56 (3): 226–240.

Smith, Philip. 2003. An Overview of the CES Seminar on "Globalization and its Impact on the World Statistical System." *Statistical Journal of the United Nations Economic Commission for Europe* 20 (2): 77–82.

Smith, Robert, László Pintér, and Charles Thrift. 2013. *Review Mechanisms of Environmental and Sustainability Indicator Systems. A Report for Environment Canada.* Winnipeg: International Institute for Sustainable Development.

Stern, Nicholas. 2007. *The Economics of Climate Change: The Stern Review*. Cambridge, UK: Cambridge University Press.

Stevens, Chantal. 2014. *Translating Indicators into Action: Data, Stories, Impact. Conference report. 2014 CIC Impact Summit.* Issaquah, WA: Community Indicators Consortium. Available at: http://www.communityindicators.net/system/publication_pdfs/89/original/2014ImpactSummitReport.pdf?1416511555.

Stiglitz, Joseph E., Amartya Sen, and Jean-Paul Fitoussi. 2008. Report by the Commission on the Measurement of Economic Performance and Social Progress. Available at: http://www.stiglitz-sen-fitoussi.fr/documents/rapport_anglais.pdf.

Swanson, Darren, and Suruchi Bhadwal, eds. 2009. *Creating Adaptive Policies: A Guide for Policy-making in an Uncertain World.* New Delhi: Sage Publications, and Ottawa: International Development Research Centre.

TEEB, The Economics of Ecosystems and Biodiversity. 2010. *The Economics of Ecosystems and Biodiversity Ecological and Economic Foundations*, ed. Pushpam Kumar. London: Earthscan.

UN, United Nations. 2007. Indicators of Sustainable Development: Guidelines and Methodologies. Third Edition. New York: United Nations. Available at: http://www.un.org/esa/sustdev/natlinfo/indicators/guidelines.pdf.

UN, United Nations. 2014a. A World That Counts. Mobilising the Data Revolution for Sustainable Development. A report by the Secretary General's Independent Expert Advisory Group on a Data Revolution. New York: United Nations.

UN, United Nations. 2014b. United Nations Secretary-General Appoints Independent Expert Advisory Group on the Data Revolution for Sustainable Development. Available at: http://www.un.org/millenniumgoals/pdf/Press%20Release_Announcement_Secretary%20General's%20Independent%20Expert%20Advisory%20Group%20on%20Data%20Revolution.pdf.

UN, United Nations. 2014c. The Road to Dignity by 2030: Ending Poverty, Transforming All Lives and Protecting the Planet. Synthesis Report of the Secretary

General on the Post-2015 Development Agenda. New York: United Nations. Available at: http://www.un.org/disabilities/documents/reports/SG_Synthesis_Report_Road_to_Dignity_by_2030.pdf.

UN, United Nations. 2015. Global Sustainable Development Report. Advance unedited version. New York: United Nations. Available at: https://sustainabledevelopment.un.org/content/documents/1758GSDR%202015%20Advance%20Unedited%20Version.pdf.

UNCSD, UN Commission on Sustainable Development Secretariat. 2012. Current Ideas on Sustainable Development Goals and Indicators. Rio 2012 Issue Briefs No. 6. Available at: https://sustainabledevelopment.un.org/content/documents/327brief6.pdf.

UN DESA, United Nations Department of Economic and Social Affairs. 2007. Indicators of Sustainable Development: Guidelines and Methodologies. New York: United Nations Department of Economic and Social Affairs.

UN DESA, United Nations Department of Economic and Social Affairs. 2013. Global Sustainable Development Report. Building the Common Future We Want. Prototype Edition. New York: UN United Nations Department of Economic and Social Affairs. Available at: https://sustainabledevelopment.un.org/content/documents/975GSDR%20Executive%20Summary.pdf.

UNDPCSD, United Nations Department for Policy Coordination and Sustainable Development. 1996. Indicators of Sustainable Development: Framework and Methodologies. New York: United Nations.

UNECE, United Nations Economic Commission for Europe. 2009. Measuring Sustainable Development. Geneva: United Nations.

UNEP, United Nations Environment Programme. 2012. Measuring Progress Towards a Green Economy. Nairobi: United Nations Environment Programme.

UN Statistical Commission. 2015. Technical report by the Bureau of the United Nations Statistical Commission on the Process of the Development of an Indicator Framework for the Goals and Targets of the Post-2015 Development Agenda. Working draft. Available at: http://unstats.un.org/unsd/broaderprogress/pdf/technical%20report%20of%20the%20unsc%20bureau%20(final).pdf.

van der Esch, Stefan, and Norav Steuer. 2014. *Comparing Public and Private Sustainability Monitoring and Reporting*. The Hague: Netherlands Environmental Assessment Agency.

van Vuuren, Detlef, and Marcel Kok, eds. 2012. *Roads from RIO+20. Pathways to Achieve Global Sustainability Goals by 2050*. The Hague: PBL Netherlands Environmental Assessment Agency.

Voß, Jan-Peter, and Arno Simons. 2014. Instrument Constituencies and the Supply Side of Policy Innovation: The Social Life of Emissions Trading. *Environmental Politics* 23 (5): 735–754.

Weber, Olaf, Roland W. Scholz, and Georg Michalik. 2010. Incorporating Sustainability Criteria into Credit Risk Management. *Business Strategy and the Environment* 19 (1): 39–50.

White, Allen L. 2012. *Redefining Value: The Future of Corporate Sustainability Ratings. Private Sector Opinion 29*. Washington, DC: International Finance Corporation.

II Learning from the Past

6 Ideas, Beliefs, and Policy Linkages: Lessons from Food, Water, and Energy Policies

Peter M. Haas and Casey Stevens

Efforts by the United Nations to draft the Sustainable Development Goals have been driven by a belief in the need for a more integrated global policy framework and to create more international communities of practice around complexes that combine diffuse issues (UNGA 2012; Jeremić and Sachs 2013). Ideas are one of the primary resources available to the United Nations (Thakur, Cooper, and English 2005; Jolly, Emmerij, and Weiss 2009). Without widespread material resources to induce behavioral change by member states, the United Nations has to fall back on the power to persuade and educate (Luck 2000; Thakur, Cooper, and English 2005). The Sustainable Development Goals, like other high-level UN declaratory initiatives, are political instruments that are intended to move the international community in a more sustainable direction by creating a powerful narrative about development to focus collective attention and action, articulating common aspirations, setting concrete goals, creating a process of learning, expanding the constituency for sustainability by building bridges between policy communities, and directing the development community's financial flows (McArthur 2013; Osborn 2013).

This chapter deductively analyzes the prospects for such sustainable issue linkage by applying Ernst B. Haas' insights about issue linkage (Haas 1980). The broad argument is that comprehensive linking of issues for a true sustainability agenda requires technical consensus about means as well as normative consensus on goals and ends. In the absence of such agreement agendas are likely to be disjointed, based on tactical linkages between smaller islands of consensus. Well-established theories exist to account for the emergence of shared norms and understandings and appraise the extent to which expert consensus about the underlying goals of sustainability currently exist at the international level, as well as the extent of understanding about the technical means of achieving them (Haas 2013).

UN efforts to couple issues have been based on two strategies: a sweeping comprehensive approach (such as discussions about a "green economy" at the 2012 UN Conference on Sustainable Development) and an incremental additive agenda (such as that pursued by the Open Working Group of the General Assembly on Sustainable Development Goals). We argue that in the absence of consensus on comprehensive sustainability goals, collective approaches to achieving sustainability through Sustainable Development Goals are likely to only occur incrementally as experts and states can reach agreement on discrete goals, their interconnections, and the policies by which they may be attained.

An ineluctable tension regards who is the likely audience for the Sustainable Development Goals: Is it governments, civil society, or the private sector? The design of Sustainable Development Goals matters depending on the primary audience. Classic conference diplomacy seeks to establish legally binding international commitments. Civil society tries to raise public consciousness, for example through the UN Non-Governmental Liaison Service Policy Briefs or the Stakeholder Forum. The private sector tries to establish voluntary benchmarks and rules of the road for global business (United Nations Global Compact 2013). The audience does matter for the greater issues regarding creating an integrative global sustainable development agenda.

We focus on states as the primary audience for the Sustainable Development Goals. Because the process of developing the Sustainable Development Goals is controlled entirely by governments through the United Nations, we presume that the audience in practice is governments and apply theories of politics about how governments engage in linkage politics.

The international community has been selectively addressing the environmental agenda since 1972 by connecting the governance of environmental issues to the governance of causally connected issues (P. M. Haas 1990, chapter 1; MacNeill, Winsemius, and Yakushiji 1991; Caldwell 1996). The governance problem is that too many problems are addressed in isolation, yielding a disarticulated and diffuse policy space. Policy networks are unable to relay information or resources for addressing and capturing externalities and synergies between issues to primary decision makers. Global governance efforts are likely to be inefficient and ineffective without taking account of the complex interconnections between issues on the agenda. This was identified as a key limitation to the effectiveness of the Millennium Development Goals, which preceded the Sustainable Development Goals. Dealing with poverty reduction, health, inequality, and environmental issues in separate silos of governance hindered efforts to deal with the agenda as a whole (UNDP and UNEP 2013).

This additive agenda has borne fruit in terms of national practices, but at the international level various declarations and endorsements over the years have not compelled widespread consensus nor significant application (Susskind, Lawrence, and Ali 2014; see also Young 2011 for a wider discussion). Many have worried that the application of the additive agenda to sustainable development might overwhelm the environment under the economic and social justice frames of those associated policy communities. For instance, the environment-security nexus has been widely questioned by civil society and academics, as has the development-environment nexus at times, because of vast asymmetries in resources between the different policy communities. Still, some increased national-level linkages, particularly in the aid area (aid and human rights, aid and gender, aid and the environment) have occurred (see Groves and Hinton 2013). The academic community started to recognize the governance challenges of tightly and loosely coupled issues within broader systems theory beginning in the late 1960s (Young 1968; Simon 1981; Perrow 1984). Particularly revealing was early work from the University of California at Berkeley, where political scientists argued that the governance of complex coupled systems depended upon the social recognition (later called the social construction) of the items to be combined (Haas 1975; LaPorte 1975; Ruggie 1975; Ruggie and Haas 1975), often due to the shared understandings of the relevant scientific community (Haas, Williams, and Babai 1977) and the international organizations with which these communities enjoy connections (E.B. Haas 1990).

Ernst B. Haas argued that the extent to which linkages are recognized and governed is a matter of intersubjective perception. The links are socially constructed, based on the persuasive power of groups advocating on their behalf (Haas 1980; Haas and Haas 1995).

Haas presents a simple matrix organized around agreement on knowledge (causal arguments) and principled norms (Haas 1980). The consensus is primarily at the level of governments and political elites, although the ideas on which political consensus is based often percolate up from domestic and transnational epistemic communities and norm entrepreneurs. Two political processes drive agenda choice: substantive and tactical linkages. Substantive linkage occurs through social learning. Tactical linkage occurs by the more traditional mechanisms of logrolling and expanding a pie to foster new coalitions.

A closer reading suggests that Haas is talking about directions of consensus, rather than absolute consensus. In fact, his model seeks to account for outcomes based on the dynamics of change among major parties. While the number of ideal types of negotiation actually becomes quite large to account for directions of change for knowledge and normative goals on different political actors, the analytic point remains valuable. With expanding goals (norms) and more consensual knowledge by experts (Haas 1980, 379–405), the policy agenda will grow to reflect the elements diplomats recognize as interconnected. Alternatively, diplomats will selectively and opportunistically combine elements according to their own domestic needs or broader geopolitical aspirations.

More recent analysis fills in the blanks about how substantive and linkage connections are made and their political foundations. The mechanisms of change for norms are norm entrepreneurs and transnational activist networks (Finnemore and Sikkink 1998; Keck and Sikkink 1998). The mechanism for change of causal beliefs are epistemic communities, either providing concordant advice to the major parties, operating through international organizations, or organized within science or expert panels (Haas 1992; Haas and Stevens 2011; Cross 2012; Haas 2012; Dunlop 2013). Causal beliefs articulated by epistemic communities are also mediated by international organizations. The causal beliefs of the epistemic community are clearly identified through their writing and through elite interviews. The degree of tolerable controversy within the community is established by the group. The relevant epistemic community has a hard core of shared beliefs and a looser belt of disagreements, although they do agree on how such disagreements may be resolved. The translation of these into policy recommendations may see more disagreement depending on the particular nature of the epistemic community.

With the presence of epistemic communities and normative consensus about the need for an expanding agenda, we expect to see the selection of Sustainable Development Goals based on their support by experts (epistemic communities) and international organizations, supported by normative arguments. This convergence of beliefs and supporting policy communities will generate a broader political process of social learning. Social learning is likely to generate more robust agendas that are capable of commanding resources from the international community and more effective outcomes as the targets reflect expert understandings. In all other cases, we expect to see tactical linkages in the absence of agreement on causal dynamics (Oye 1979; Sebenius 1983). Such goals will contribute to broader learning about the connections between elements in a complex global agenda. Tactical linkages are unlikely to spawn broader learning processes and are unlikely to be politically resilient. Table 6.1 presents the prospects for issue linkage about sustainability in various normative and epistemic arrangements, and the political processes by which issues would emerge on the international agenda.

In this chapter we look for whether there is articulation of consensus and particular aspects of contestation on issues related to the agenda of the Sustainable Development Goals. For analysis, we see evidence of consensus by major interlocutors in contexts where discussions across different international forums largely cohere on problem definition and general approach for addressing the problem. Based on the discussion above, this is not about absolute consensus and the end of contestation, but rather contexts where norm entrepreneurs or epistemic communities have subordinated contestation or have compartmentalized it sufficiently to allow consensus to become powerful. For our analysis, the focus is largely on the international level and primarily in international organizations because this is most relevant to the creation of the Sustainable Development Goals. It is possible that the observed consensus at this level does not correspond to the consensus and contestation at other levels. It is

Table 6.1
Prospects for Issue Linkage

Normative Consensus / Causal Consensus	Normative consensus is converging around interconnected goals through norm entrepreneurs and transnational action networks	Normative consensus is not converging around interconnected goals in the absence of norm entrepreneurs and transnational action networks
Consensus on causal ideas is converging through an epistemic community and independent international organizations	Social learning Issues: Water, food security, energy security	Incremental tactical linkage
Consensus on causal ideas is not converging in the absence of an epistemic community	Incremental tactical linkage Issues: gender equality, human rights, peaceful societies, rule of law, sustainable consumption and production	Incremental tactical linkage Issues: education, sustainable cities, transportation, employment, information communication and technology

worth noting again that this may be directions of consensus rather than absolute consensus and that consensus may be narrowly construed by the actors involved.

There do not appear to be any higher-order norms under which the Sustainable Development Goals can be nested that command widespread consensus. Even the commitments under the UN Charter and responsibility for UN intervention and compliance with the Millennium Development Goals remain contested and subject to varied interpretation (Doyle 2011). Common but differentiated responsibilities remain limited to just the climate convention and the Kyoto Protocol, and efforts to extend this to the larger agenda are quite contested in the negotiations on forming the Sustainable Development Goals. "Sustainable development" itself was a tactical agreement by the Brundtland Commission Report (Timberlake 1989; Haas 1996), and current discussions about sustainable development are riven by an abiding conflict over North-South financial commitments. The links between economic development, environmental protection, human rights, security, and justice command consensus in certain limited circles, but there is little causal consensus about how to govern the linkages. "Planetary boundaries" are contested within the scientific community and lack complete linkages as to how to manage social activities to avoid exceeding such global limits.

Environmental protection also lacks universal or even necessarily majoritarian normative support, since many of the major international environmental treaties lack the ratification or support of at least one country that would play a significant role in resolving the problem at hand (Iwama 1992; Sand 1992; Choucri, Sundgren, and Haas 1994; Beyerlin 2007; Sands and Peel 2012). The treaties on ozone-depleting substances are somewhat aberrant in this respect, and efforts to deal with climate change, biodiversity loss, desertification, and precautionary trade agreements all have major states remaining outside. Human rights may be such a higher-order norm that could support the Sustainable Development Goals and has been argued to be foundational at the 2012 UN Conference on Sustainable Development and the negotiations that followed. However, the human rights and sustainable development principles are not yet sufficiently fully developed to provide normative coherence at this point (Darrow 2012).

Young and colleagues' proposal for a *Grundnorm* (this volume, chapter 3) would provide the opportunity for longer-term linkages for the successors to the Sustainable Development Goals. In the absence of agreement on the causal processes, such a norm would enhance the possibility for tactical linkages between different agenda items.

The experience of the Millennium Development Goals, by and large, seems to confirm this theoretical approach. Poverty reduction was successfully addressed because it did reflect common norms and causal understandings (Fukuda-Parr 2011; Fukuda-Parr and Hulme 2011). Some health goals were largely achieved, based on common causal beliefs and normative consensus (Murray, Frenk, and Evans 2007; Arregoces et al. 2012). Other components of the Millennium Development Goals have been less successful. The Millennium Development Goals offer a cautionary tale for the Sustainable Development Goals. They fostered suspicion by the UN member states because of the way the indicators and benchmarks for the Millennium Development Goals were developed by the Secretariat independently of governments. Because of this experience, states have been careful to control the outcomes of the process during the two years of negotiations in the UN General Assembly and through the Open Working Group.

Causal Consensus, Norms, and the Sustainable Development Goals

What, then, are the prospects for Sustainable Development Goals according to this theoretical framework? In the absence of any universal beliefs about universal interconnections between issues, the development of Sustainable Development Goals had to rely on identifying consensus behind individual goals, rather than a top-down process based on the presumptive need for a comprehensive package of issues. We consider individual issue areas that are likely to satisfy both the causal and normative requirements used in our approach. Over time, we hope that focusing on such individual topics may encourage policy-relevant learning about the interconnections between them.

We consider a wide universe of candidates for the Sustainable Development Goals developed from "The Future We Want" from the 2012 UN Conference on Sustainable Development, the Expert Group Meeting on Science and Sustainable Development report, the Open Working Group agenda, and other issues that have been raised in this process. This includes a wide range of issues, some of which build from the Millennium Development Goals, some which expand upon them, and others that are a combination of issues.

In the preparation for the 2012 UN Conference on Sustainable Development, delegates considered the "green economy" and the transformation to low- or zero-carbon industrial systems one of the fundamental goals of the conference. Ultimately, the issue failed to command strong support because of North-South disagreement about financing and the distribution of benefits (Barbier 2011; Barbier 2012; Bina 2013). The result has been that the discussions on Sustainable Development Goals have taken an inclusive approach toward issues rather than organizing them around a central direction for reducing poverty and increasing sustainability. This was highlighted in the Open Working Group outcome, which worked for most of its time with 17 different goals.

The discussions around Sustainable Development Goals were modified by forum shopping by epistemic communities and normative entrepreneurs. Parallel discussions about these issues in the United Nations, in environmental treaty regimes, and in other international forums like the World Trade Organization and World Economic Forum may have been preferred areas for discussion by many of the actors. While epistemic coherence on the Sustainable Development Goals agenda may be quite limited, this may be the result of the expenditure of persuasive resources in other forums.

Analyzing consensus across issue areas is difficult, and assertion of consensus may be deployed by actors to reduce space for other actors to contest the efforts. As justified above, our approach to measuring consensus largely focused on increasing or decreasing consensus at the international level for the period 2010–2015. Assessing the coherence of the discussions across various forums such as the 2012 UN Conference on Sustainable Development, the World Business Council on Sustainable Development, and the World Economic Forum can provide broad understanding of increasing causal and normative consensus. The limitation of this approach is that it may misidentify consensus where actors have stopped contesting but deep divides remain, and that contestation for some of these issues might be most relevant at other levels than the international. However, the functional consensus measured here may be most relevant when ascertaining the processes that gave rise to the Sustainable Development Goals, since it identifies the intersubjective understanding of the actors involved.

Table 6.1 shows the array of prominent issues around Sustainable Development Goals in the theoretical approach explained in the last section. The implementation of the Sustainable Development Goals in the 2015–2020 period could transform many of these placements, but this identifies their epistemic and normative basis for much of the shaping process.[1]

Areas of Low Consensus

Clear causal and normative bases simply are absent on issues like education and urban sustainability. This can be the result of significant uncertainty about causal relationships, partial causal consensus, and core debates about epistemological and normative principles. Such distinctions are demonstrated in the case of education discussions. While education is one of the first issue areas to have developed goals and targets at the international level, these remain broad and focused on issues of wide agreement with significant areas undiscussed (Beatty and Pritchett 2012). While reports from the UN Educational, Scientific and Cultural Organization (UNESCO) and other organizations have probed into these other issues, the governance targets developed at conferences at Jomtien in 1990, Dakar in 2000, and integrated into the Millennium Development Goals have all remained focused on simply universal access to education and reducing gender discrepancies (Goldstein 2004; King 2007). While UNESCO's Education for All coalition has worked on these issues, clear guidance remains limited (UNESCO and UN Children's Fund 2013). Information communication and technology may be an ideal case of these issues where a focus on technical aspects related to the specific field has dominated the agenda and thus limited efforts to develop shared causal and normative understandings for interactions with other aspects of sustainable development. Environment and development actors call for interaction with information communication and technology issues, but the consensus on how to interact with these issues is not clear. Other issues, such as urban sustainability and sustainable transportation, have not seen causal or normative consensus largely because of rifts within their fields more generally (for the example from urban sustainability, see Williams 2010).

Climate change, desertification and land erosion, and public health (broadly) have developed causal or normative consensus, but the extension of this to the rest of the sustainable development agenda is limited at this point. Most clearly, climate change has seen significant efforts to develop fundamental causal understanding, policy-usable knowledge, and norm dynamics, but the application to the sustainable development agenda is uncertain. While some health issues saw significant progress under the Millennium Development Goals, expansion to public health more generally has more significant contestation (Haffner and Shiffman 2013). However, the connection to other issues in the post-2015 agenda are in a far murkier area of climate change adaptation (IPCC 2012). Desertification and land erosion, some health issues, and even ocean issues have similar dynamics. The development of causal understanding within the individual issue areas does not guarantee interlinkage understanding; however, these issue areas may be likely to see transformation as the Sustainable Development Goals evolve and tactical linkages give rise to higher-order social dynamics. This process we will explore in the final section.

Gender equality, human rights, peaceful societies, and governance issues have seen significant investment by state and nonstate actors in constructing a firm normative basis for action, but this far outpaces causal consensus. While norm pushers have been key in highlighting these issues, the presentation for sustainability often takes the form of silver bullet or "add-and-stir" arguments that underplay the specifics of interconnection with various sustainability issues. Human rights, for instance, has significant actors pushing for its inclusion in the agenda on Sustainable Development Goals, with a clear focus on individual personal liberties. This group includes prominent nongovernmental organizations, international organizations, and governments. However, the articulation of connection and interactions is rarely based upon some shared causal consensus, but rather is often a silver bullet articulation of human rights as essential to every other issue.

In discussions of Sustainable Development Goals, these issues are often treated without setting priorities or recognizing more important or valuable connections; instead, they are pushed for with universal integration across issues (an issue as true with the Sustainable Development Goals as it was with the Millennium Development Goals; see Alston 2005). Gender equity is certainly developing a more robust understanding of causal connections to various issues (OECD 2010); however, there remain different approaches to dealing with these connections in practice (Chant and Sweetman 2012). Sustainable consumption and production form a distinct issue area for understanding because there are firm normative communities on some specific issues (such as ending fossil-fuel subsidies and fertilizer management), but sustainable consumption and production lacks an overall normative understanding. Efforts like the Green Economy framework by the UN Environment Programme have provided some shared understanding about the issue and a partial framework, but not a fully developed understanding.

Areas of High Consensus

The areas with the most significant development of causal consensus and normative consensus are food security and nutrition, water, and energy security. We apply these concepts simply, as they were invoked by the authors and expert groups we have surveyed.

The extent of consensus is partially a result of deliberate efforts by actors to build new connections about the linkages between discrete issues.[2]

However, some consensus was simply based on prior understandings by actors within the different issue areas of the clear interrelationships between the issues. None of these three issue areas has achieved full causal consensus, but all three have seen some consensus reduce contestation.

Food security saw increasing consensus around efforts to link rural agriculturalists into larger supply chains following the food price spike of 2007–2008. Significant contestation remains regarding loans to such rural agriculturalists, the use of genetically modified organisms, and other important issues; however, consensus has improved. Water issues have similarly seen consensus around the broad framework of integrated water resource management. Earlier debates about neoliberalism of water resources and the scale of water management were of decreasing importance as integrated water resource management achieved prominence on the international agenda (Mukhtarov and Cherp 2015). Significant implementation and operationalization consensus remains elusive within integrated water resource management. Energy security issues maintain a consensus around nonintervention in the global energy market while striving to gradually decouple emissions from economic production. However, nuclear issues remain prominent in this story.

Case Studies

We now analyze in more detail the conditions that have led to the consensus around these three—food security, energy security, and water—where consensus has increased in recent years.

Food Security

Food issues are illustrative of the conditions of increasing consensus and the potential impacts from this consensus. While there remain significant debates and discussion on significant parts of the agenda, the normative frame of "food security" and a shared focus on supply-chain issues by various groups has increased consensus since the food price spike of 2007–2008. We focus significantly on this case as it highlights issues relevant to water and energy, which will be discussed below.

There was little shared understanding on food security in the years before the 2007–2008 global spike in food prices. The World Economic Forum reports in the mid-2000s did not emphasize small-scale agriculturalists, instead emphasizing large agricultural development at the national levels. The focus solely on global agricultural trade has spurred an active opposition to the agenda, centralized in the "food sovereignty" efforts. The food price spike of 2007–2008, tied with the larger financial crisis and a spike in energy prices, largely "caught the world by surprise" (Karapinar 2010, 1; see also Piesse and Thirtle 2009). The free market approach of the World Economic Forum early in the 2000s became significantly challenged by a variety of different actors (McKeon 2014). While there remains significant contestation on many points—notably around issues like sustainable agriculture, access to credit for farmers, and genetically modified organisms—some degree of consensus has formed around a supply-chain view of food security that emphasizes local landholders. The policy problem is seen as one of inefficient and unproductive supply-chains, but the solution is seen in enabling local agriculturalists to tie into these supply chains. This consensus is seen with the increasing coherence between the UN system focused on hunger-related issues and the organizations with a focus on agricultural trade liberalization (Margulis 2013).

The global spike in food prices was a surprise to most governments and international actors because "during the previous two decades real prices had fallen, and there was unusually low price volatility around the declining trend" (Jayasuriya, Mudbhary, and Broca 2012, vii). While prices did not return to the level of the last major world food crisis of 1973–1975, prices in basic grains began increasing in August 2006 and significantly increased, with rice prices presenting the most significant global spike (Dawe and Slayton 2011). While causation of the food crisis was complex, many policy makers at the national level responded to the rising food crisis with around 40 countries imposing bans or restrictions on exports of food (Mittal 2009). The food shock, the surprise to policy makers, and these national responses all combined to generate the conditions for institutional reassessment at the global level. While prices stabilized briefly, another increase followed in 2010–2011.

For many decades, the United Nations had a number of different institutions focused on issues of hunger and agricultural production more generally. This includes the UN Food and Agriculture Organization (FAO), the World Food Programme, the International Fund for Agricultural Development, and others. These institutions, and particularly the FAO and the World Food Programme, have a prominent role in hunger and agricultural policies around the world, significant staff, and long histories dealing with hunger issues. However, "the UN itself was surprisingly ill-prepared for the 2007-2008 price spike policy crisis. No major conference on food insecurity was in the pipeline" (Lang and Barling 2012, 314). The institutions had normative focus on the problems and broad efforts to feed the hungry and boost agricultural production in the medium term, but there did not appear to be a clear consensus on the long-term governance pathways toward food security. Without ongoing efforts to organize an international response and without a toolbox of policy options to provide to states, the UN system began efforts to respond largely in the early part of 2008. The UN System High-level Task Force on the Global Food Security Crisis was established in April, and a bio-energy conference scheduled for June was repackaged to deal with food security in general (Lang and Barling 2012).

The most significant institutional transformation within the UN system was the reform of the Committee on World Food Security. While it was established in the 1970s food price crisis, for much of its existence the committee "has played a relatively minor role in international politics and was generally ineffective and inactive due to a lack of interest and buy-in from member states and insufficient budget" (Duncan and Barling 2012, 147). Deciding in 2008 to undertake significant reform, negotiations between the bureau of the Committee on World Food Security and various stakeholders resulted in a significantly reformed institution by 2009. The institution was strengthened, meetings become more prominent, and connections with civil society were expanded. Food chains and supply systems were not emphasized as a key part of the food security agenda at the time of this reform; however, by 2011 the issues of food chains figured prominently in the discussion. An important aspect of some of these discussions, in connection to the main contentions of this chapter, is that they were often cast as an interlinkage between issues of energy security and food security, with renewables being seen as key in linking rural agriculturalists in developing countries into the global food-supply systems.

The World Trade Organization was similarly ill-prepared to deal with the food shock, since much of the basis for the decade of negotiations on agricultural trade had occurred based on the stability of prices. In addition, the insulation policies adopted by a number of WTO members to protect their domestic food resources were seen as a new challenge in pushing for the core principles of the institution (Karapinar 2010). Scheduled WTO meetings in July 2008 focused significantly on the issue of agricultural trade, but failed to reach substantial agreement. At the meeting, the surge in food prices was used simultaneously for those pressing for quicker finalization of a trade liberalization agenda and those pushing for maintaining national policy opportunities to deal with such shocks. The food price spike of 2007–2008 thus introduced significant problems for the WTO ongoing negotiations. The focus then switched to developing approaches that could "[reassure developing countries] about the world market being an affordable source of food supplies" (Konandreas 2011). The UN Special Rapporteur on the Right to Food called for a "compatibility check" between WTO rules and the global efforts to address the food price spike. Of particular emphasis was improving the speed and reach of global food-supply systems. This is emphasized in a least-developed countries organized policy reflection at the July 2008 WTO meeting on the food crisis. The central question for the discussion was whether WTO rules impede responses to the crisis or can provide a start on solutions to the food crisis. Delegates at this meeting questioned structural adjustment programs, the focus on biofuels, and the exclusion of many developing countries from regional and global food-supply chains. While trade liberalization remains the overall objective of the organization, efforts to connect with this supply-chain focus on food security had begun to be seen as a WTO-rules-based response to the price spike.

Governance extended beyond these two institutions with other notable efforts. The Group of 20 major economies took up the issue in 2011, and their first action plan focused on increasing agricultural production in developing countries and increased market transparency (Clapp and Murphy 2013). While the Group of 20 discussions have had a significant influence on the agenda and may even have crowded out other efforts (Clapp and Murphy 2013), their focus on supply-chain aspects of food security is largely an outgrowth of discussions within the UN system and the WTO. The issue is similar in other organizations that have taken on efforts related to food security, like the UN Commission on Trade and Development and the Organisation for Economic Co-operation and Development. Taking their lead from the WTO and the UN system, these efforts similarly focused on the supply-chain issues following the price spike.

Food security has been a major topic of discussion for decades now. There were already 25 articles dealing with "food security" in the Web of Knowledge by 1990, and it became an organizational aspect for international discussions about hunger and malnourishment in the 1990s. The Rome Declaration on World Food Security, agreed to in 1996, focused on "achieving food security for all and to an ongoing effort to eradicate hunger in all countries, with an immediate view to reducing the number of undernourished people to half their present level no later than 2015" (FAO 1996). This goal would eventually form the core of the Millennium Development Goal focused on hunger and nutrition. The focus on food security in the Rome Declaration on World Food Security emphasizes three different aspects: "availability of staple foods, stability of supplies, and access for all to these supplies" (Mechlem 2004). The NGO forum at the meeting challenged this core agenda by encouraging a greater focus on agrarian reform to support smallholder agriculturalists and questioned global economic structures, most notably a governance system built around intellectual property rights, intensive use of industrial agrochemicals, and the use of genetically modified organisms. The first challenge was organized by a number of nongovernmental organizations around a rival conceptual framework of "food sovereignty," initiated at the La Via Campesina Tlaxcala Conference in April 1996. Focused on rural livelihoods, this questions the global market focus of the Rome Declaration. While the Rome Declaration was largely agnostic about the use of biotechnology and genetically modified organizations to address hunger and malnutrition problems, this would become a major contentious aspect in scientific and governance discussions following the Rome meeting. The FAO established the FAO Biotechnology Forum to deal with the debate around biotechnology, and at their first conference in 2000 on the issue the "polarization" of views about biotechnology and hunger problems took center stage. The debate about this takes many shapes, with different countries (notably the United States and Europe; see Falkner 2007), different international organizations, and even the NGO community emphasizing biotechnology as a solution to production in developing countries or emphasizing regulation and caution in the application of such technology.

However, a focus on food-supply chains and the particular methodologies deployed by various institutions has carved out a specific focus shaping global institutions. The increasing food pressures to develop over the next decades may very well require "a revolution in the social and natural sciences concerned with food production" (Godfray et al. 2010). The political institutions, particularly the Committee on World Food Security and its connections to civil society (Clapp and Murphy 2013), offer opportunities for expanded efforts on food security issues and may be well placed to deal with interactive sustainability aspects.

Water

The epistemic consensus around water issues is fairly limited to holistic water management, most notably integrated water resource management, and a conceptualization of water in security terms. These two ideas are not mutually exclusive in practice (Cook and Bakker 2012), but their combination together has had significant impact on global policy-making discussions. Holistic water management has a long international tradition, going back at least to the UN Conference on Water in 1977. Although many actors were experimenting with holistic water management in the 1980s and early 1990s, the frame of integrated water resource management really became important with the foundation of the World Water Council in the mid-1990s. This nongovernmental think tank organized a number of international events to push integrated water resource management and some specific policies, such as full water pricing and improved supply systems. In six World Water Forums so far, the World Water Council has brought together various stakeholders to continually call for adoption and refinement of integrated water resource management (Rahaman and Varis 2005). The World Water Council was not alone in calling for holistic water management in the particular form of integrated water resource management, but they were a major agenda setter. Following their second World Water Forum in 2000, the number of scientific publications increased significantly, and businesses and nongovernmental organizations began deploying integrated water resource management in a number of projects.

Major problems remained in forming normative consensus around water issues, with key divisions between actors on the approach to privatization of water resources and dams in the 1990s and early 2000s. In addition, although integrated water resource management has been recognized as important by a number of governments, implementation proved difficult, and many actors found the articulation of holistic policy to be highly problematic (Pahl-Wostl et al. 2011; Pahl-Wostl et al. 2013). Despite these problems, integrated water resource management has become "the 'lingua franca' of global water governance" and has "played a role in smoothing up a number of sensitive conflicts in the area of water resources, such as the debate on neo-liberalization of water governance, the debate over the scale at which water resources are managed best, and about the roles and responsibilities of various policy actors" (Mukhtarov and Cherp 2015, 9). Integrated water resource management appears to have a consensus about broad terms of discussion rather than a fully developed policy relevant approach. The definition remains vague, and efforts are focused more on getting actors to agree to the framework than developing specific policy-relevant knowledge.

While integrated water resource management had coalesced in the early 2000s, the normative shape of water security would only come later in the decade. Starting with the World Business Council for Sustainable Development's 2002 report "Water for the Poor," the business and economic community began focusing on and formulating an agenda for water security (Goldman 2007). The debate became further coordinated when the World Economic Forum took up the issue in 2008. The World Bank and the

International Monetary Fund were similarly developing policy frames around water security at this point, but a clear international agenda was still quite controversial. Privatization and corporate ownership of water supply generated significant controversy at national and international levels. Focused on supporting technological innovation and better water pricing, the food security coalition organized between businesses and other actors had begun to develop the broad outlines of a norm of decentralized water governance (World Economic Forum 2011). As the 2013 UN-Water report on water security explains, "policies are needed on water planning, allocation and pricing, aimed at increasing water security through increased water efficiency in industrial, agricultural and domestic water use" (UN-Water 2013). The groups organizing around integrated water resource management (such as UN-Water, UNESCO's water groups, etc.) and the groups organized around water security (namely the World Economic Forum and the Global Water Partnership) utilized increasing attention on water as a result of the 2005–2015 Decade of Action on Water for Life and the 2013 International Year of Water Cooperation to create far more focus on these core issues. These ongoing processes created an overlap between the two communities. Multiple reports and events emphasized water security as part of an integrated, holistic approach to managing water. Core debates remain within the large umbrella of integrated water resource management, notably over operationalization and community involvement in management and different views of water resources spanning a human rights approach to an economic commodity approach (Gupta and Nilsson, this volume, chapter 12).

Energy Security

The core of energy security arguments is a belief that markets and technological innovation will be the crucial parts of decreasing energy-supply vulnerability. This is the result of increasing complexity in global energy supply (with the addition of natural gas to traditional oil price problems), the initial failure for decentralized energy-supply systems to provide reliable power to all, and the decrease of competing international institutions governing energy supply (namely OPEC and the International Energy Agency; Cherp and Jewell 2011). These processes have given rise to advanced market models for states to understand their energy security and a variety of indicators to measure vulnerability and insecurity (Cherp and Jewell 2013).

In the context of these general processes, application of energy security to the green economy framework has focused on decreasing energy market volatility, expanding advanced electricity supply, and trying to reduce carbon output. International organizations have promoted energy efficiency and energy transitions as key in efforts to decouple economic output from energy input. The 2013 International Energy Agency report specified that its "findings show there is significant potential to decouple economic growth—and energy production and use, in particular—from its proven environmental impacts" (International Energy Agency 2013). Similarly, at a prominent side event at the 2012 UN Conference on Sustainable Development on "Decoupling for Change," the UN Energy Group emphasized efficiency and decoupling in many different reports, and the UN Sustainable Energy for All project is focused primarily on decoupled energy efforts and energy efficiency. Smart grid application, integration of renewable energy into the mix, and general implementation of efficiency are supported as ways to stabilize energy supply and demand relationships. The agreement on creating systems that encourage what the World Economic Forum calls "smart globalization" and improved trade relationships in fuels, technology, and eventually engineering knowledge, may form a coherent normative and epistemological goal of these various international organizations.

While aiming to reduce price volatility, largely trusting the market to accomplish a long-term transition, the nuts and bolts of this process remain quite contested. This is most clearly exemplified with nuclear power as an option for energy security. Right before the Fukushima nuclear disaster in Japan in 2010, the International Energy Agency supported a resurgence of nuclear energy investment to contribute to carbon reductions. The early years of the International Renewable Energy Agency have been beset by significant contestation of their approach to nuclear issues, eventually leading to the head of the agency dismissing nuclear energy as not within their scope (Van de Graaf 2013).

While economic knowledge about supply and vulnerability indicators rooted in these models has become increasingly advanced, and international organizations increasingly support technological, market-led solutions, it is possible that this is not exactly reflective of the epistemological and normative consensus that our framework would expect to be essential. For example, this may be a case where normative agreement between powerful actors creates the image of epistemological consensus. The dominance of economic models in energy systems and the prominence of important institutional actors may prevent challenges from being clearly seen. However, the policy use of this work may create a functional epistemological consensus that yields similar results.

Dynamics and Interactions for a Cascading Sustainability Agenda

Food security, water, and energy security not only have significant coherence in their own policy communities, but also potential policy linkages to other issues. Notably, with focus on the "nexus" in the past few years, including in the World Economic Forum, the UN Department of Economic and Social Affairs, and separate Nexus conferences in 2011 and 2014 organized by the Stockholm Environment Institute, the interaction between the three issues has received significant attention. Right now there is no nexus community, however, and interactions remain rooted in one of the different policy communities. The terrain remains defined by different issue islands.

However, some of these issues and connections may be expanded through social learning as the process develops. The most likely issue here is the interaction between food security and water security, both because the issue areas are closely connected and because the communities have begun significant interactions over the past few years. While energy is often included with these other issues, the connection appears limited to biomass production and the impact it has on food security, with only limited broader engagement.

Other connections can be made between these issues and others that lack sufficient conditions for social learning. Energy security and sustainable consumption and production could see synergies related to fossil-fuel subsidies and mining. Similarly, food security and issues of sustainable cities and employment offer significant possibilities for connections. A policy space may exist focused on rural employment, which bridges all three of these areas. This would require some transition from the normative and epistemological focus of food security currently (maybe integrating some of the zero-hunger normative aspects), but there is space for it, and some initial conceptual articulations that may yield results.

In addition to substantive learning between issues, there are opportunities for more traditional issue-linkage and logrolling from these issues. Water, energy, and food issues all offer opportunities for linking funding and technological transfer with policy changes. While food security was initially discussed in 2008 as an attempt to generate a reinvestment in rural areas, little consistent support materialized across countries (McMichael and Schneider 2011). While learning between food security and rural employment is possible, for example, more likely is traditional bargaining. Similarly, recognitions of connections between water and ecosystem issues (including biodiversity and oceans) may generate consensus as integrated water resource management gets increasing funding and refinement. Similarly, water discussions have emphasized governance issues and adopted a human rights focus consistently in negotiations of the Sustainable Development Goals. Treatment of water, energy, and food by the World Economic Forum and the World Business Council for Sustainable Development demonstrate some of the tactical linkages between these three issues. Efforts to expand technology that solves all three issues is often the focus of the discussions, and tradeoffs are minimized in efforts to deal with various problems (see World Business Council for Sustainable Development 2009). At this point, the various discussions about the nexus between water, food, and energy in the World Economic Forum, World Business Council for Sustainable Development, and elsewhere remain largely tactical linkages and not substantive consensus around interlinked issues. This is also reflected in the discussions that formed the Sustainable Development Goals in the United Nations. While there were some connections made and a focus on interlinkages was key, those for food, water, and energy remained significantly limited in their holistic vision. In the final list of goals, they appear as largely separate efforts with limited interconnection.

Further institutional designs through the Sustainable Development Goals and the High-level Political Forum may contribute to the capacity building for longer-term consensus about the development and sustainability agenda (Bernstein, this volume, chapter 9). As states and policy networks become more experienced with the governance of individual items on the sustainability agenda, they may come to recognize the causal linkages between those issues and others, leading to a more comprehensive and intertwined agenda in the future.

The learning dynamics around Sustainable Development Goals are different from those of the Millennium Development Goals, despite the heavy reliance on these goals as a framework for the Sustainable Development Goals. The Millennium Development Goals generated some ad hoc and partial learning in the areas of public health and poverty alleviation, but little in terms of more integrated global policy vision. The Sustainable Development Goals have lacked the intellectual impulse behind the Millennium Development Goals. The Millennium Project, the primary science network for the Millennium Development Goals, only operated in the early years of the program. Assessment of the project was constantly beset by out-of-date information that was often not synchronous with information about other goals.

Conclusion

Goal-based governance efforts have shown limited impacts to transform the epistemic and normative deliberations for the Sustainable Development Goals. The context for the Sustainable Development Goals to shape the international policy environment is one where there is no clear epistemic and normative basis for sustainable development. The analysis above argues that on many of the issues in the Sustainable Development Goals, consensus on either normative or epistemic issues has not been increasing in recent years. This does not make issue linkage impossible, but it does mean that it is likely to be incremental. In contrast, the analysis found that increasing consensus in recent years in the international policy communities on water, food, and energy offers some potential for social learning. While dissent and disagreement exist on these issues at other levels, at the international policy level this disagreement has been somewhat limited in recent years. The Sustainable Development Goals will take place in this environment, and recognition of the issue linkage context will be important for the goal-setting activity to achieve the largest impact on sustainability.

The prospects for social learning in the framework of Sustainable Development Goals are likely to be limited at the outset to those of food security, energy security, and water issues. The other issue areas are likely to result in traditional bargaining over goals and will have less advanced targets and indicators as a result. However, there is some space in the efforts to link the various problems in the Sustainable Development Goals and in the creations of multi-issue indicators and traditional logrolling between issues to allow these to spread to other areas like nutrition, employment, and the wider environmental agenda.

In the longer term, learning about connections between some of the nexus issues may inform a more comprehensive set of post-2030 Sustainable Development Goals. Further discussions about these areas of consensus, and their implications for linkages to other issue areas, may help generate a broader and more integrated agenda for the future, as well as the foundations for a new *Grundnorm*.

Notes

1. Some issues in the Sustainable Development Goals are not included in table 6.1, since actors and institutions approach the issues in very different ways. This is the case with issues of peace/justice, oceans, and means of implementation. While

aspects of these may see consensus, the issue as discussed at the international level with the Sustainable Development Goals was too wide-ranging. Our approach would expect incremental, tactical linkages on parts of these issues at best.

2. A clear example of these efforts is in the various nexus-themed conferences. The ideas of linking food security, energy, and water issues together in global governance efforts started in the late 1990s with the rise of discussions of "virtual water" or the importation of agricultural goods to replace domestic agricultural use. The nexus became a significant part of the agenda with the Bonn 2011 conference. Developed from discussions at the World Economic Forum and aiming to influence the 2012 UN Conference on Sustainable Development, the German government brought together a number of stakeholders to more fully develop the concept for governance. A number of important meetings followed this initial focus on water, food, and energy in a shared policy response, many focused around the World Decade on Water Action. Climate has often been included in the conversations recently.

References

Alston, Philip. 2005. Ships Passing in the Night: The Current State of the Human Rights and Development Debate Seen Through the Lens of the Millennium Development Goals. *Human Rights Quarterly* 27 (3): 755–829.

Arregoces, Leonardo, Felicity Daly, Catherine Pitt, Justine Hsu, Melisa Martinez-Alvarez, Giulia Greco, Anne Mills, et al. 2012. Countdown to 2015: Changes in Official Development Assistance to Maternal, Newborn, and Child Health in 2009–10, and Assessment of Progress Since 2003. *Lancet* 380: 1157–1168.

Barbier, Edward. 2011. The Policy Challenges for Green Economy and Sustainable Development. *Natural Resources Forum* 35: 233–245.

Barbier, Edward B. 2012. The Green Economy Post Rio+20. *Science* 338: 886–887.

Beatty, Amanda, and Lant Pritchett. 2012. From Schooling Goals to Learning Goals: How Fast Can Student Learning Improve? *CGD Policy Paper* 12.

Beyerlin, Ulrich. 2007. Different Types of Norms in International Environmental Law. In *The Oxford Handbook of International Environmental Law*, ed. Daniel Bodansky, Jutta Brunnée and Ellen Hey, 426–448. Oxford: Oxford University Press.

Bina, Olivia. 2013. The Green Economy and Sustainable Development: An Uneasy Balance? *Environment and Planning. C, Government & Policy* 31 (3): 1023–1047.

Caldwell, Lynton Keith. 1996. *International Environmental Policy*. Durham, NC: Duke University Press.

Chant, Sylvia, and Caroline Sweetman. 2012. Fixing Women or Fixing the World? "Smart Economics," Efficiency Approaches, and Gender Equality in Development. *Gender and Development* 20 (3): 517–529.

Cherp, Aleh, and Jessica Jewell. 2011. The Three Perspectives on Energy Security: Intellectual History, Disciplinary Roots and the Potential for Integration. *Current Opinion in Environmental Sustainability* 3 (4): 202–212.

Cherp, Aleh, and Jessica Jewell. 2013. Energy Security Assessment Framework and three Case Studies. In *International Handbook of Energy Security*, ed. Hugh Dyer and Maria Julia Trombetta, 146–173. Northampton, MA: Edward Elgar Publishing.

Choucri, Nazli, Jan Sundren, and Peter M. Haas. 1994. More Global Treaties. *Nature* 367 (3): 405.

Clapp, Jennifer, and Sophia Murphy. 2013. The G20 and Food Security: A Mismatch in Global Governance? *Global Policy* 4 (2): 129–138.

Cook, Christina, and Karen Bakker. 2012. Water Security: Debating an Emerging Paradigm. *Global Environmental Change* 22 (1): 94–102.

Darrow, Mac. 2012. The Millennium Development Goals: Milestones or Millstones-Human Rights Priorities for the Post-2015 Development Agenda. *Yale Human Rights and Development Law Journal* 15: 55–128.

Davis Cross, Mai'a K. 2012. Rethinking Epistemic Communities Twenty Years Later. *Review of International Studies* 39 (1): 137–160.

Dawe, David, and Tom Slayton. 2011. The World Rice Market in 2007–08. In *Safeguarding Food Security in Volatile Global Markets*, ed. Adam Prakash, 171–182. Rome: Food and Agriculture Organization.

Doyle, Michael W. 2011. Dialectics of a Global Constitution: The Struggle Over the UN Charter. *European Journal of International Relations* 18 (4): 601–624.

Duncan, Jessica, and David Barling. 2012. Renewal Through Participation in Global Food Security Governance: Implementing the International Food Security and Nutrition Civil Society Mechanism to the Committee on World Food Security. *International Journal of Sociology of Agriculture and Food* 19 (2): 143–161.

Dunlop, Claire A. 2013. Epistemic Communities. In *Routledge Handbook of Public Policy*, eds. Eduardo Araral, Scott Fritzen, Michael Howlett, M. Ramesh, and Xun Wu, 229–243. New York: Routledge.

Falkner, Robert, ed. 2007. *The International Politics of Genetically Modified Food: Diplomacy, Trade, and Law*. Basingstoke, UK: Palgrave Macmillan.

FAO, Food and Agriculture Organization. 1996. Declaration on World Food Security. World Food Summit. Rome: Food and Agriculture Organization.

Finnemore, Martha, and Kathryn Sikkink. 1998. International Norm Dynamics and Political Change. *International Organization* 52 (4): 887–917.

Fukuda-Parr, Sakiko. 2011. Theory and Policy in International Development. *International Studies Review* 13: 122–132.

Fukuda-Parr, Sakiko, and David Hulme. 2011. International Norm Dynamics and the "End of Poverty.". *Global Governance* 17: 17–36.

Godfray, H. Charles J., John R. Beddington, Ian R. Crute, Lawrence Haddad, David Lawrence, James F. Muir, Jules Pretty, et al. 2010. Food Security: The Challenge of Feeding 9 Billion People. *Science* 327 (5967): 812–818.

Goldman, Michael. 2007. How "Water for All!" Policy Became Hegemonic: The Power of the World Bank and Its Transnational Policy Networks. *Geoforum* 38 (5): 786–800.

Goldstein, Harvey. 2004. Education for All: The Globalization of Learning Targets. *Comparative Education* 40 (1): 7–15.

Groves, Leslie, and Rachel Hinton, eds. 2013. *Inclusive Aid: Changing Power and Relationships in International Development*. New York: Routledge.

Haas, Ernst B. 1975. Is There a Hole in the Whole? *International Organization* 29 (3): 827–876.

Haas, Ernst B. 1980. Why Collaborate? Issue Linkage and International Regimes. *World Politics* 32 (3): 357–405.

Haas, Ernst B. 1990. *When Knowledge is Power: Three Models of Change in International Organizations*. Berkeley: University of California Press.

Haas, Ernst B., Mary Pat Williams, and Don Babai. 1977. *Scientists and World Order*. Berkeley: University of California Press.

Haas, Peter M. 1990. *Saving the Mediterranean*. New York: Columbia University Press.

Haas, Peter M. 1992. Introduction: Epistemic Communities and International Policy Coordination. *International Organization* 46 (1): 1–37.

Haas, Peter M. 1996. Is Sustainable Development Politically Sustainable? *Brown Journal of World Affairs* 3 (2): 239–248.

Haas, Peter M. 2012. Epistemic Communities. In *The Oxford Companion to Comparative Politics*, ed. Joel Krieger, 351–359. Oxford: Oxford University Press.

Haas, Peter M., and Ernst B. Haas. 1995. Learning to Learn: Improving Global Governance. *Global Governance* 1 (3): 255–284.

Haas, Peter M., and Casey Stevens. 2011. Organized Science, Usable Knowledge and Multilateral Environmental Governance. In *Governing the Air*, ed. Rolf Lidskog and Göran Sundqvist, 125–161. Cambridge, MA: MIT Press.

Hafner, Tamara, and Jeremy Shiffman. 2013. The Emergence of Global Attention to Health Systems Strengthening. *Health Policy and Planning* 28 (1): 41–50.

International Energy Agency. 2013. World Energy Outlook. Paris: Organisation for Economic Co-operation and Development and International Energy Agency.

IPCC, Intergovernmental Panel on Climate Change. 2012. *Managing the Risks of Extreme Events and Disasters to Advance Climate Change Adaptation*. Cambridge, UK: Cambridge University Press.

Iwama, Toru. 1992. Emerging Principles and Rules for the Prevention and Mitigation of Environmental Harm. In *Environmental Change and International Law*, ed. Edith Brown Weiss, 107–123. Tokyo: UNU Press.

Jayasuriya, Sisira, Purushottam Mudbhary, and Sumiter Singh Broca. 2013. *Food Price Spikes, Increasing Volatility and Global Economic Shocks: Coping with Challenges to Food Security in Asia*. Rome: Food and Agriculture Organization.

Jeremić, Vuk, and Jeffrey D. Sachs. 2013. The United Nations in the Age of Sustainable Development. UN High-level Advisory Panel. New York: General Assembly of the United Nations.

Jolly, Richard, Louis Emmerij, and Thomas G. Weiss. 2009. *UN Ideas that Changed the World*. Bloomington: Indiana University Press.

Karapinar, Baris. 2010. Introduction: Food Crises and the WTO. In *Food Crises and the WTO: World Trade Forum*, eds. Baris Karapinar and Christian Häberli, 1–22. London: Cambridge University Press.

Keck, Margaret E., and Kathryn Sikkink. 1998. *Activists Beyond Borders*. Ithaca: Cornell University Press.

King, Kenneth. 2007. Multilateral Agencies in the Construction of the Global Agenda on Education. *Comparative Education* 43 (3): 377–391.

Konandreas, Panos. 2011. Global Governance: International Policy Considerations. In *Safeguarding Food Security in Volatile Global Markets*, ed. Adam Prakash, 329–360. Rome: Food and Agriculture Organization.

Lang, Tim, and David Barling. 2012. Food Security and Food Sustainability: Reformulating the Debate. *Geographical Journal* 178 (4): 313–326.

LaPorte, Todd R., ed. 1975. *Organized Social Complexity*. Princeton: Princeton University Press.

Luck, Edward C. 2000. Blue Ribbon Power: Independent Commissions and UN Reform. *International Studies Perspectives* 1 (1): 89–104.

MacNeill, Jim, Pieter Winsemius, and Taizo Yakushiji. 1991. *Beyond Interdependence: The Meshing of the World's Economy and the Earth's Ecology*. New York: Oxford University Press.

March, James G., and Johan P. Olson. 1988. The Institutional Dynamics of International Political Orders. *International Organization* 52 (4): 943–970.

Margulis, Matias E. 2013. The Regime Complex for Food Security: Implications for the Global Hunger Challenge. *Global Governance: A Review of Multilateralism and International Organizations* 19 (1): 53–67.

McArthur, John W. 2013. Own the Goals: What the Millennium Development Goals Have Accomplished. *Foreign Affairs* 92: 152–162.

McKeon, Nora. 2014. *Food Security Governance: Empowering Communities, Regulating Corporations*. New York: Routledge.

McMichael, Philip, and M. Schneider. 2011. Food Security Politics and the Millennium Development Goals. *Third World Quarterly* 32 (1): 119–139.

Mechlem, Kerstin. 2004. Food Security and the Right to Food in the Discourse of the United Nations. *European Law Journal* 10 (5): 631–648.

Mittal, Anuradha. 2009. *The 2008 Food Price Crisis: Rethinking Food Security Policies*. New York: United Nations.

Mukhtarov, Farhad, and Aleh Cherp. 2015. The Hegemony of Integrated Water Resources Management as a Global Water Discourse. In *River Basin Management in the Twenty-First Century: Understanding People and Place*, eds. Victor Roy Squires, Hugh Martin Milner and Katherine Anne Daniell, 3–21. Boca Raton, FL: CRC Press.

Murray, Christopher J. L., Julio Frenk, and Timothy Evans. 2007. The Global Campaign for the Health MDGs: Challenges, Opportunities, and the Imperative of Shared Learning. *Lancet* 370 (9592): 1018–1020.

OECD, Organisation for Economic Co-operation and Development. 2010. Gender Inequality and the MDGs: What Are the Missing Dimensions? Paris: OECD.

Osborn, Derek. 2013. Building on Rio+20 to Spur Action for Sustainable Development. *Environment* 55 (3): 3–13.

Oye, Kenneth A. 1979. The Domain of Choice. In *The Eagle Entangled*, ed. Kenneth A. Oye, Donald S. Rothchild and Robert J. Lieber, 3–33. New York: Longman.

Pahl-Wostl, Claudia, Ken Conca, Annika Kramer, Josefina Maestu, and Falk Schmidt. 2013. Missing Links in Global Water Governance: A Processes-Oriented Analysis. *Ecology and Society* 18 (2): 1–10.

Pahl-Wostl, Claudia, Paul Jeffrey, Nicola Isendahl, and Marcela Brugnach. 2011. Maturing the New Water Management Paradigm: Progressing from Aspiration to Practice. *Water Resources Management* 25 (3): 837–856.

Perrow, Charles. 1984. *Normal Accidents: Living With High Risk Technologies*. New York: Basic Books.

Piesse, Jenifer, and Colin Thirtle. 2009. Three Bubbles and a Panic: An Explanatory Review of Recent Food Commodity Price Events. *Food Policy* 34 (2): 119–129.

Rahaman, Muhammad Mizanur, and Olli Varis. 2005. Integrated Water Resources Management: Evolution, Prospects and Future Challenges. *Sustainability: Science. Practice and Policy* 1 (1): 15–21.

Ruggie, John Gerard. 1975. International Responses to Technology. *International Organization* 29 (3): 557–584.

Ruggie, John Gerard, and Ernst B. Haas. 1975. International Responses to Technology. *International Organization* 29 (3): 557–583.

Sand, Peter H., ed. 1992. *The Effectiveness of International Environmental Agreements*. Oxford: Grotius.

Sands, Philippe, and Jacqueline Peel. 2012. *Principles of International Environmental Law*. Cambridge, UK: Cambridge University Press.

Sebenius, James K. 1983. Negotiation Arithmetic. *International Organization* 37 (2): 281–316.

Simon, Herbert A. 1981. *The Sciences of the Artificial*. Cambridge, MA: MIT Press.

Susskind, Lawrence E., and Saleem H. Ali. 2014. *Environmental Diplomacy: Negotiating More Effective Global Agreements*. Oxford: Oxford University Press.

Thakur, Ramesh Chandra, Andrew Fenton Cooper, and John English, eds. 2005. *International Commissions and the Power of Ideas*. Tokyo: United Nations University Press.

Timberlake, Lloyd. 1989. The Role of Scientific Knowledge in Drawing up the Brundtland Report. In *International Resource Management: The Role of Science and Politics*, ed. Steinar Andresen and Willy Ostreng, 117–123. London: Belhaven Press.

UNDP and UNEP, UN Development Programme and UN Environment Programme. 2013. Breaking Down the Silos: Integrating Environmental Sustainability in the Post-2015 Agenda. New York: Thematic Consultation on Environmental Sustainability in the Post-2015 Agenda.

UNESCO, UN Organization for Education, Science and Culture, and UN Children's Fund. 2013. Making Education a Priority in the Post-2015 Development Agenda. Paris: UN Organization for Education, Science and Culture.

UNGA, United Nations General Assembly. 2012. The Future We Want. UN Doc. A/RES/66/288.

United Nations Global Compact. 2013. Report to the United Nations Secretary-General: Corporate Sustainability and the United Nations Post-2015 Development Agenda. New York: United Nations Global Compact.

UN-Water. 2013. Water Security and the Global Water Agenda. Hamilton, Ontario: United Nations University-Institute for Water, Environment and Health.

Van de Graaf, Thijs. 2013. Fragmentation in Global Energy Governance: Explaining the Creation of IRENA. *Global Environmental Politics* 13 (3): 14–33.

Williams, Katie. 2010. Sustainable Cities: Research and Practice Challenges. *International Journal of Urban Sustainable Development* 1 (1–2): 128–132.

World Business Council for Sustainable Development. 2009. Water, Energy and Climate Change: A Contribution from the Business Community. Geneva: World Business Council for Sustainable Development.

World Economic Forum Water Initiative. 2011. *Water Security: The Water-Food-Energy-Climate Nexus*. Washington, DC: Island Press.

Young, Oran R. 1968. *A Systemic Approach to International Politics*. Princeton, NJ: Center of International Studies, Woodrow Wilson School of Public and International Affairs, Princeton University Press.

Young, Oran R. 2011. Effectiveness of International Environmental Regimes: Existing Knowledge, Cutting-edge Themes, and Research Strategies. *Proceedings of the National Academy of Sciences of the United States of America* 108 (5): 19853–19860.

7 Lessons from the Health-Related Millennium Development Goals

Steinar Andresen and Masahiko Iguchi

When assessing the Sustainable Development Goals, it makes sense to consider the lessons learned from the Millennium Development Goals. While the former are universal goals and the latter focuses on developing countries, there are still similarities between these two goal-based approaches (see Young, this volume, chapter 2). What are the successes and failures of the Millennium Development Goals, and how can the best lessons be applied while avoiding the pitfalls and mistakes?

In the next section of this chapter we go through the status of the Millennium Development Goals and account for what goals have been achieved and where goal achievements are less impressive. While a lot has been achieved, there has also been considerable criticism and questions raised as to their significance. Maybe the most fundamental criticism is the difficulty in establishing a causal link between the Millennium Development Goals and actual performance on the ground. In other words, examination of the significance of "externalities" to the success of the Millennium Development Goals is critical to draw implications for the Sustainable Development Goals. This crucial point has not been much addressed in UN circles.

Our account illustrates that some goals have been met, and others have not; these are categorized in table 7.2 below as achievements and remaining challenges. Considering that goals may be quite random and the strong variations in underlying causal factors, this does not tell the whole story in terms of what efforts have been initiated and achievements made in reaching the various goals. We zoom in on this aspect in the third section, focusing on the health-related goals. They have played a very prominent role in the overall focus of the Millennium Development Goals, since three of the eight goals are addressing health issues. Our main focus is on Goal 4, which addresses reducing mortality among children. We will illustrate that a massive effort has been invested by various types of actors to reach this goal, and it is particularly important to study, as at least some of these efforts can be causally linked to the Millennium Development Goals. To substantiate this, we will first focus on the Global Alliance for Vaccines and Immunization (GAVI). Then we describe and discuss the role played by Norway in this context, since the health-related Millennium Development Goals have been very high on the political agenda in Norway. This illustrates that leadership of various kinds can make a crucial difference in coming closer to realizing the UN goals (Young 1991; Underdal 1994).

In the final section, we discuss what lessons can be learned from the Millennium Development Goals in general, and the health-related goals in particular, in designing the Sustainable Development Goals.

Background of the Millennium Development Goals

This section provides the basic background of the Millennium Development Goals. To this end, first, it describes the origins of Millennium Development Goals by focusing on how, by whom, when, and why these goals were formulated. Then, it reviews the achievements and failures of the Millennium Development Goals by looking at existing literature in this field.

According to Hulme (2009), the formulation of the Millennium Development Goals can be divided into the following periods: post-World War II in the 1940s to the key UN summits in the 1970s; International Development Goals proposed by the Development Assistance Committee at the Organization for Economic Co-operation and Development (OECD) during 1990s; and the series of discussions toward the UN Millennium Summit in the late 1990s to 2000 to amend International Development Goals into the Millennium Development Goals in 2001.

The origin of the Millennium Development Goals can be traced back to Article 25.1 in the UN Declaration of Human Rights of 1948, which states that "everyone has the right to a standard of living adequate for the health and well-being of himself and his family, including food, housing and medical care" (UNGA 1948). Based on this principle, in the 1970s the UN resolution "International Development Strategy for the Second United Nations Development Decade" set out the objective for each developed country to "progressively increase its official development assistance to the developing countries and exert its best efforts to reach a minimum net amount of 0.7% of its gross national product at market prices by the middle of the decade" (UNGA 1970).

Needs to establish concrete development goals were raised during the 1990s in order to counter the decline of international aid. According to Hulme and Fukuda-Parr (2011), this political and ideological motivation led to promoting international development as a global project. To this end, the Development Assistance Committee proposed a set of International Development Goals in 1996 with the following three dimensions: economic well-being (measures to combat poverty), social development (including education, gender, and health), and environmental sustainability and regeneration (see table 7.1 for details).

However, the International Development Goals did not have much practical impact in many of the OECD countries, given that they lacked a plan of action. Hence, the goals only gained media attention for a few days. It is also important to point out that this document was produced entirely by rich countries to come up with a list of achievable, concrete, and measurable goals that would appeal to OECD members, and it is not surprising that "the document's premise and promotion of 'partnership' sounded like standard aid agency rhetoric" among poor countries (Hulme 2009, 17).

Important initiatives that influenced the creation of Millennium Development Goals were the Human Development initiative of the UN Development Programme (UNDP) and the World Bank's income and poverty-monitoring initiative (Saith 2006). In particular, while the International Development Goals placed higher priority on economic growth and poverty reduction, the UNDP's Human Development Report 1997 focused on human development goals based on a human rights approach, such as life expectancy, disease eradication, and adult literacy (UNDP 1997).

Two years later, the United Nations started to globally set targets by making plans for the Millennium Assembly of the United Nations in 1998, under leadership of the new UN Secretary-General, Kofi Annan, who was keen to make global poverty reduction central to the UN agenda. Accordingly, the key document, "We the Peoples: The Role of the United Nations in the 21st Century" (as known as the "Millennium Report") was launched in April 2000 by Kofi Annan as the basic document to work toward the Millennium Summit. Along with poverty reduction, it emphasized the importance of gender equality and women's empowerment, reproductive health, health issues including combatting HIV/AIDS, economic growth, access to new technology including information technology, social development, and the environment, and global partnerships to enhance development assistance (UN 2000). While many of these goals were reflected in the

Table 7.1

Comparisons of International Development Goals and Millennium Development Goals

International Development Goals	Millennium Development Goals
1. Economic wellbeing: The proportion of people living in extreme poverty in developing countries should be reduced by at least one-half by 2015 2. Social development: There should be substantial progress in primary education, gender equity, basic health care, and family planning as follows: - Universal primary education in all countries by 2015. - Progress toward gender equity disparity in primary and secondary education by 2005. - The death rate for infants and children under the age of 5 years should be reduced in each developing country by two-thirds the 1990 level by 2015. The rate of maternal mortality should be reduced by three-fourths during this same period. - Access should be available through the primary health care system to reproductive health service for all individuals of appropriate ages, including safe and reliable family planning methods, as soon as possible and no later than the year 2015. 3. Environmental sustainability and regeneration: There should be a current national strategy for sustainable development, in the process of being implemented, in every country by 2005 to ensure that current trends in the loss of environmental resources—forests, fisheries, fresh water, climate, solid, biodiversity, stratospheric ozone, the accumulation of hazardous substances, and other major indicators—are effectively reversed at both global and national levels by 2015.	1. Eradicate extreme hunger and poverty: - Halve, between 1990 and 2015, the proportion of people whose income is less than $1 a day - Halve, between 1990 and 2015, the proportion of people who suffer from hunger. 2. Achieve universal primary education: - Ensure that, by 2015, children everywhere, boys and girls alike, will be able to complete a full course of primary schooling. 3. Promote gender equality and empower women: - Eliminate gender disparity in primary and secondary education, preferably by 2005, and in all levels of education no later than 2015. 4. Reduce child mortality: - Reduce by two-thirds, between 1990 and 2015, the under-5 mortality rate. 5. Improve maternal health: - Reduce by three-quarters, between 1990 and 2015, the maternal mortality ratio. 6. Combat HIV/HIDS, malaria, and other diseases: - Have halted by 2015 and begun to reverse the spread of HIV/AIDS - Have halted by 2015 and begun to reverse the incidence of malaria and other major diseases. 7. Ensure environmental sustainability: - Integrate the principles of sustainable development into country policies and programs and reverse the loss of environmental resources

Table 7.1 (continued)

International Development Goals	Millennium Development Goals
	- Halve, by 2015, the proportion of people without sustainable access to safe drinking water and basic sanitation - Have achieved by 2020 a significant improvement in the lives of at least 100 million slum dwellers. 8. Develop a global partnership for development: - Develop further an open, rule-based, predictable, nondiscriminatory trading and financial system - Address the special needs of the least-developed countries - Address the special needs of landlocked developing countries and small-island developing states - Deal comprehensively with the debt problems of developing countries through national and international measures in order to make debt sustainable in the long term - In cooperation with developing countries, develop and implement strategies for decent and productive work for youth - In cooperation with pharmaceutical companies, provide access to affordable essential drugs in developing countries - In cooperation with the private sector, make available the benefits of new technologies, especially information and communications.

Source: Authors based on Development Assistance Committee (1996, 9–11) and UN (n.d.).

Millennium Declaration adopted in the Millennium Summit of the United Nations in September 2000, health-related goals were largely expanded in the declaration due to strong lobbying from health and gender equality lobbyists and support from the World Bank, World Health Organization, and UN Children's Fund (Hulme 2009). As a result, health-related goals occupy three out of eight of the Millennium Development Goals (see table 7.1 for details). Given the critical importance of health-related goals, their evaluations are dealt in the following section.

Many scholars have assessed achievements and remaining challenges of the Millennium Development Goals and drawn lessons for the post-2015 development agenda (see table 7.2). While the Millennium Development Goals have led to many achievements, this section argues that under-achievement of Millennium Development Goals can be understood by Young's notion of "fit," which concerns how institutions are designed to "match" problems and their solutions (Young 2002).

Among the major achievements, existing studies evaluate the success of the Millennium Development Goals as follows. First, since the Millennium Development Goals placed their primary focus on poverty eradication and increasing international development aid, they promoted improvements in poverty eradication as well as increased financial aid from institutions and official development assistance and raised the priority of policies relating to poverty eradication in developing countries (Manning 2010; Moss 2010; Vandermoortele 2011). Importantly, Millennium Development Goals did so by changing the norms and the discourse on development, shaping ideological perspectives (Fukuda-Parr 2010). Second, they enhanced sectoral linkages among several sectors, such as health and water quality, sanitation and nutrition, and so forth (Vandermoortele 2011), and promoted the participation of many stakeholders in a number of developing countries (Langford 2010).

On the other hand, remaining challenges of the Millennium Development Goals are partially explained by the following factors. First, they lack interlinkages from global to national and local levels (Katsuma 2008), and the manifestations of the gaps vary significantly among countries. For example, Africa as a whole made much less progress on the goals than Asia due to the domestic interethnic and communal conflicts, as well as the focus of the Millennium Development Goals on social service, not infrastructure (Agwu 2011; Peterson 2010). Furthermore, Easterly (2009) points out the Millennium Development Goals are "unfair" to Africa—for instance, Goal 4 on child mortality is based on *proportional terms* and not on *absolute terms*; therefore, it was difficult for Africa to meet this goal, since it is the highest-mortality region in the world.

Second, they do not reflect the needs of recipients in the regional context (Shepherd 2008). To put it simply, the Millennium Development Goals did specify an overall goal, but they did not set out a specific process to make it possible with reference to national priorities (Fukuda-Parr 2010). This leads to another shortcoming pointed out by many: The goals lacked implementation mechanisms, especially in terms of financing, where they placed an excessive focus on donor funding (Clemens, Kenny, and Moss 2007). Moreover, Saith (2006) claims that while the goals are a "wish list," the targets are set out at a more detailed level based on indicators with a time frame. Hence, it is questionable whether the Millennium Development Goals can function as a "programming tool" to deliver the targets they set.

The third broad category of criticism addresses the nature of the targets. Because the Millennium Development Goals were formulated based on the idea of results-based management, it was difficult to measure goals and targets such as human rights, equality, or even the question of "good governance," which were not included (Alston 2005; Hulme 2007; Nelson 2007; Vandemoortele and Delamonica 2010). Furthermore, with regard to environmental sustainability, many important global environmental issues such as climate change and protection of biodiversity were not explicitly mentioned.

The final and most profound criticism is the question whether any improvements associating with Millennium Development Goals are the direct result of the Millennium Development Goals, or whether results are linked primarily to externalities. For example, rapid democratization and radical technological innovations can significantly enhance a country's economic prosperity, and hence would contribute to improvements of many Millennium Development Goals. Conversely, political corruptions or radical changes in political structure may significantly delay progress.

This last point, the issue of externalities, or the question of causality (Young 2008), is at the core of the evaluation of the Millennium Development Goals. In other words, a closer examination of factors establishing a causal link between the Millennium Development Goals and the relevant problems is needed to draw lessons from the Millennium Development Goals. For example, it was not because of the Millennium Development Goals that China has brought 400 million of its people out of poverty, but due to the conscious policies and priorities of the Chinese government.

Table 7.2

Achievements and Remaining Challenges of the Millennium Development Goals

Goals	Achievements	Remaining challenges
Goal 1: Eradicate poverty and hunger	- Proportion of people living on less than $1.25 a day dropped to 14% in 2015. - Proportion of undernourished people fell to 12.9% in 2014-2016.	- About 800 million people still live in extreme poverty.
Goal 2: Achieve universal primary education	- Net rate of primary school enrollment reached 91% in 2015. - Literacy rate of youth increased from 83% in 1990 to 91% in 2015.	- 57 million children are not in school. - Children from the poorest households are four times as likely to be out of school as children from the richest households in developing regions.
Goal 3: Gender equality	- The developing regions as a whole achieved the target to eliminate gender disparity in primary, secondary, and tertiary education.	- Women are more likely to live in poverty than men. - Women earn 24% less than men globally.
Goal 4: Reduce child mortality	- Reduction rate of child mortality tripled globally since 1990.	- About 16,000 children die each day before the age of 5. - Child mortality rate of the poorest households is almost twice that of the richest households in developing regions.
Goal 5: Improve maternal health	- Maternal mortality rate has declined by 45% worldwide since 1990.	- Maternal mortality rate in developing regions is 14 times higher than in developed regions.
Goal 6: Combat HIV/AIDS, malaria, and other disease	- New HIV infections fell from an estimated 3.5 million cases to 2.1 million between 2000 and 2013. - Over 6.2 million malaria deaths have been averted between 2000 and 2015.	- About 36% of the 31.5 million people living with HIV in developing regions received antiretroviral therapy in 2013.
Goal 7: Ensure environmental sustainability	- 91% of the global population is using an improved drinking water source in 2015.	- About 16% of the rural population does not use improved drinking water sources in developing regions. - Global carbon dioxide emissions have increased by over 50% since 1990.

Table 7.2 (continued)

Goals	Achievements	Remaining challenges
Goal 8: Global partnership for development	- Official development assistance from developed countries increased by 66% between 2000 to 2014, reaching $135.2 billion.	- Only Demark, Luxembourg, Norway, Sweden, and the United Kingdom continued to exceed the official development assistance target of 0.7% of gross national income.

Source: Authors based on UN (2015).

Health-related Goals: Achievements and Challenges

This section focuses on one sector addressed by the Millennium Development Goals: health policies. In somewhat simplified terms, the health-related Millennium Development Goals are as follows: Goal 4 seeks to reduce the child mortality rate by two-thirds for children under five years between 1990 and 2015. Goal 5 seeks to improve maternal health and reduce by three-quarters between 1990 and 2015 the maternal mortality rate and achieve universal access to reproductive health. Goal 6 targets HIV/AIDS, malaria, and other diseases, and sought to achieve by 2010 universal access to HIV medicine and to have halved victims by 2015 (with similar goals for malaria and other diseases).

Some analysts are very critical of the use of targets, especially regarding whether we have data sufficiently good to measure what has been achieved. Particularly, they point to the difficulty of getting good and reliable data for the 1990 baseline (Attaran 2005). This criticism may well be valid, and, if so, it probably applies to many if not most of the quantitative Millennium Development Goals. However, as we are not in a position to judge and evaluate this controversy, in the following we have chosen to disregard this methodological uncertainty and use official figures from the World Health Organization (WHO 2013).

As for Goal 4, significant progress has been made in reducing child mortality: In 2012, 6.6 million children under five years of age died, compared to 12.6 million children in 1990. Thus, child mortality has been reduced by almost 50%. The global rate of decline has also accelerated recently. The rate of immunization coverage has increased considerably, as two-thirds of the WHO members have reached at least 90% coverage. However, the score is much lower in Sub-Saharan Africa. Despite considerable progress, the goal will not be reached by 2015.

Regarding Goal 5, despite significant reductions in the number of maternal deaths, the decline is less than half what is needed to reach the stated goal. Also, less than two-thirds in the relevant age used contraception. Goal achievement here is therefore considerably lower than for the two other health-related goals.

As for Goal 6, in 2012 the number of newly infected was down by one-third compared to 2001, but the Sub-Sahara accounted for more than 70% of those infected. Somewhat paradoxically, the number of people living with HIV will continue to grow as they live longer due to better treatment; there were some 35 million people living with the disease in 2012, an increase from previous years. Considerable progress has also been made in terms of reducing incidence of malaria and tuberculosis, but progress has (again) been less in Sub-Saharan Africa.

However, large geographical areas will reach the health-related goals without foreign aid playing a significant role. This applies to the whole of Latin America, China, North Africa, and Southeast Asia. The major challenges are sub-Saharan Africa, Pakistan, and large parts of India, and the main bottleneck is malfunctioning domestic health systems.

Some of these positive results are probably to some extent causally linked to the Millennium Development Goals, while others would have happened in any case. However, according to observers that have followed global health politics more closely than we have: "The [health-related] Millennium Development Goals have inspired, mobilized, involved and not the least facilitated the funding of key activities on a scale that the world has not previously experienced" (Lie et al. 2011). Key actors involved in achieving the health-related goals are large parts of the UN family as well as the World Bank, a number of governments, and "not the least the medical journal *The Lancet* with editor Richard Horton at the helm since 2003 has issued a range of series dealing with health-related goals. Time and again strong criticism of global injustice has been voiced and focus directed on the tragic loss of almost nine million lives every year due to poverty. The series have undoubtedly contributed to a debate that continues to rise in intensity" (Lie et al. 2011). The role of the media is important regarding health-related goals, particularly the child-related goal. TV reports show dramatic pictures of suffering children and more direct access to quick information reveals global injustice. The Child Convention, adopted in 1989, has been ratified by all but two countries and may also strengthen this goal.

In terms of resources, development assistance from public and private sources targeting maternal and child health has increased by some 400% in recent years. However, development assistance channeled through the UN system declined in the same period (McNeill, Andresen, and Sandberg 2013). That is, government assistance outside the United Nations—and not least private capital—was behind this astonishing development. Of particular importance as well is the establishment of various types of partnerships. One such partnership is the WHO's Partnership for Maternal, Newborn and Child Health. Its main task is lobbying and providing knowledge for effective interventions, and it has 400 members from public and nongovernmental organizations. Another one is the "Count Down" initiative, a loosely knit constellation that arranges large conferences every second year to monitor the extent to which various countries are on track to meet Goals 4 and 5.

Of particular importance are the kinds of partnerships that have been instrumental in generating new funds, such as the Global Fund to Fight Aids, Tuberculosis and Malaria as well as GAVI, which we shall zoom in on below.

As we will demonstrate in the next two sections on the role of GAVI and Norway in realizing this Millennium Development Goal, various types of leadership have been particularly important. While the role played by GAVI is a typical example of instrumental leadership, the role played by Norway qualifies as leadership by example. However, by diving somewhat deeper into the material we see that leadership by *individuals*, an aspect often overlooked in the literature, looms large in both cases. This individual leadership also overlaps in the two cases, and it is illustrated by high-level representatives of political offices, international organizations, and private business. The network may also qualify as a small but powerful epistemic community.

The Role of Partnerships: The Case of the Global Alliance for Vaccines and Immunization

GAVI was *not* created as a result of the adoption of the Millennium Development Goals in 2000. Nevertheless, its creation in 1999 was given added momentum by the process around the Millennium Development Goals, since there was a surge in interest and initiatives on global health around the turn of the millennium. Our main concern here is the extent to which this innovative initiative has contributed to some of the results achieved under this goal. We will therefore first give a brief account of results that have been delivered by GAVI. Second, we will discuss the nature of this institution, focusing on the role of various actors as well as individuals.

Does it represent an approach that can be set up in other areas as well in order to achieve better results under goal-based UN governance?

The following account is based on the second independent evaluation of GAVI conducted in 2010 (GAVI Board 2010). Seen from the aggregate macro level, it appears that GAVI has delivered impressive results: It has provided a total of US$2.2 billion in disbursements to 75 countries over the period 2000–2009. The WHO estimates that GAVI's vaccine support averted nearly four million future deaths before 2009 (GAVI Board 2010). Needless to say, methodological uncertainties and disputes regarding the figures abound. According to the WHO's evaluation report, however, even taking into account a substantial margin of error, GAVI's work has been a very significant achievement. In terms of added value, of high and direct relevance to Millennium Development Goal 4, it is concluded: "There is good evidence to suggest that GAVI has been able to attract additional funding for immunization and its major donors would not have contributed to the immunization on the scale that they did without it" (GAVI Board 2010, 7). GAVI has also played a key role in getting the issue high on the international political agenda and has been instrumental in securing additional resources to vaccination through its various mechanisms in relation to the Millennium Development Goals, particularly to Millennium Development Goal 4. In short, GAVI has probably been the most important mechanism to reach a fairly high goal achievement here.

Turning, then, to the questions of how and why GAVI was established and what can account for its significant achievements, the following is a brief and simplified account of this complex and fascinating story (drawing on Muraskin 2005; Sandberg and Andresen 2010; McNeill, Andresen, and Sandberg 2013; McNeill and Sandberg 2014).

Since its creation in 1948, the WHO was the undisputed global leader in terms of health issues until the mid-1990s, as a result of its high UN-based legitimacy as well as long-standing expertise in the field. This started to change in the 1990s for at least two reasons. Its authority and effectiveness in terms of delivering new and necessary vaccines had been strongly reduced, not least because of its strained relationship with the pharmaceutical industry. Secondly, new actors entered the scene at the time, not least the more powerful World Bank as well as influential nonstate actors like the Program for Appropriate Technology in Health, challenging the dominant role of the WHO. The late 1990s were therefore characterized by turf battles and lack of cooperation between established partners like the WHO and UN Children's Fund and new actors like the World Bank and innovative nongovernmental organizations.

These parties met in working groups in 1998 and started discussions to address the need to launch new initiatives to speed up vaccination, but little progress was made. Then, somewhat out of the blue, Bill Gates emerged on the scene. Based on the detailed account given by Muraskin (2005), the fact that he became interested in vaccination was completely accidental. However, when de did, he wanted an effective small body outside the UN system. He also chose to focus on vaccination, as this was seen as a measureable and cost-effective way to speed up vaccination and thereby save lives. Negotiations started between the above mentioned actors and, initially, the Program for Appropriate Technology in Health (also based in Seattle), who first represented Bill Gates and what subsequently came to be known as the Bill and Melinda Gates Foundation.

Initially, the WHO and the UN Children's Fund were cautious, favoring a UN basis and wary that a newcomer could undermine their role in this area. In 1999, a compromise emerged through the establishment of GAVI, by which the established UN bodies as well as the Program for Appropriate Technology in Health became key partners based on their comparative competences. It has been noted that some key individuals were crucial in forging compromise, for example the WHO Director-General Gro Harlem Brundtland and one of her key advisors, Mr. Tore Godal. The pharmaceutical industry was also brought on board thanks to the more pragmatic approach by Brundtland, as opposed to her predecessor in the WHO. Important as their roles may have been, however, this initiative may have gone nowhere had it not been for the money provided by the Gates Foundation, contributing the initial funding of no less than US$750 million. As has been pointed out, providing a lot of money makes it considerably easier for previous competitors to come to agreement (McNeill, Andresen, and Sandberg 2013). The Gates Foundation also contributes to a number of other institutions, for example the Global Fund, and is thereby instrumental to achieving other Millennium Development Goals as well.

Over time, the GAVI Secretariat has been strengthened at the expense of the partners, who are now considered more as subcontractors. Opinions vary over the significance of this development, but it is outside the scope of this article to discuss this. The important point here is that GAVI has continued to provide impressive results. Two points of criticism, however, may be worth considering in our context. One is that GAVI is too preoccupied with its numerical cost-effective approach and is neglecting broader issues that are more difficult to deal with, as well as to quantify and measure results. This point may have some merit, and the criticism also has bearing upon the Millennium Development Goals more generally. The other criticism is more uncommon in the discussion of the merits and shortcomings of global governance: GAVI is too strong and dominant and gives too little say to recipients. This point is also relevant, and it is interesting in the sense that, usually, weak global institutions reduce the effectiveness of global governance; it is seldom a problem that they are too strong.

Contributing to the impressive results of GAVI are also the strong commitments of key donor states, and interestingly the United States is by far the largest donor. Another significant player has been Norway, an actor with a very high priority to reach the health-related Millennium Development Goals.

The Role of Active Countries: The Case of Norway

In addition to partnerships, individual states have also played an important role in global health governance. One example we discuss here in more detail is Norway. Norway is among the few countries in the world that has long committed more than 0.7% in official development assistance. In fact, Norway is very close to the top among all countries, committing almost 1% of gross domestic product. In addition, global health has long figured prominently both as a component of development aid as well as an important aspect of Norway's more general foreign policy. Norway was important in establishing and funding the "old" health institutions like WHO and the UN Children's Fund, and more recently GAVI and the Global Fund. Norway was also one of the main supporters behind the establishment of the Millennium Development Goals, not least the health-related goals.

This was also motivated by the fact that Brundtland had just become Director-General of the WHO. She was previously a long-standing Norwegian prime minister, a medical doctor, and a well-known figure in the international arena through her leadership of the process toward *Our Common Future*. This made her well suited for that position. Norway quickly became a key champion of these goals, and the effort has lasted and expanded over time. This was strengthened by the role of key individuals. Brundtland had brought Jonas Gahr Støre, her pervious chief of staff at the prime minister's office, to Geneva as her head of staff, thereby introducing him to the field of global health. He brought this experience with him when he became minister of foreign affairs in 2005, a position he kept until 2012. In 2005, the new social democratic government singled out global health and the vaccination of children as an issue of high priority. Mr. Støre was also eager to underline the key position of health in foreign policy. This perspective was behind the initiative to establish the Oslo Ministerial Group in 2007, a meeting forum for seven developed and developing countries to push the role of health higher on the international agenda (Møgedal and Alveberg 2010). The work of GAVI and Bill Gates on vaccination also struck a chord with Prime Minister Jens Stoltenberg. The focus on the cost-effective health benefits of vaccination mirrored the thinking of the economist Stoltenberg, who had long underlined the gospel of cost-effectiveness in Norway's climate policy.

In 2007 he launched the Global Campaign for the Health Millennium Development Goals and an associated network of global leaders. To follow up, in 2009 he announced in the United Nations that, until 2020, Norway would contribute 3 billion Norwegian kroner, or close to US$400 million, to global cooperation for female and children's health. Norway has thereby been among the most significant funders of efforts to improve global health; its funding has tripled since 1990, making Norway also among the largest donors in absolute terms (Norwegian Ministry of Foreign Affairs 2011). This applies to GAVI, the Global Fund, and the UN Children's Fund, as well as the WHO. Because the maternal health goal has been the furthest from being reached, more financing has been channeled to improve performance there. Bilateral partnerships have also been established with four key developing countries—India, Pakistan, Nigeria, and Tanzania—to support these countries' domestic programs to improve maternal and child health.

The significance of individuals is also illustrated by the role played by Godal, a key architect behind GAVI. He has been a very important player in Norwegian global health policy as a senior advisor (Sandberg and Andresen 2010). This activist position has been continued by the new conservative government since 2013, as witnessed by the appointment of Norwegian Prime Minister Erna Solberg as UN ambassador to oversee the progress toward the Millennium Development Goals.

According to the last government white paper on the issue, "Health in Foreign and Development Policy": "Norway shall through political leadership, diplomacy and economic support be active in mobilizing for a strong global consensus for cooperation to take care of national health needs" (our translation, Norwegian Ministry of Foreign Affairs 2011, 5). To mobilize for women and children's rights was singled out as the government's highest priority—referring explicitly to the three health-related Millennium Development Goals. The gist of the argument, reflecting the work of Brundtland and her team in the WHO, was that good health was seen as one key condition for development and reduction of poverty. The human rights perspective is also essential in the Norwegian engagement for realizing the Millennium Development Goals.

Not surprisingly, Norway also played a very active role in realizing the Sustainable Development Goals. Instead of detailing the efforts made by Norway, we will discuss briefly what signals the Sustainable Development Goals send to Norway, underlining the difference between the Millennium Development Goals and the Sustainable Development Goals. While the Millennium Development Goals had an explicit Southern assistance perspective, the Sustainable Development Goals also point to severe challenges in the North if the goal of a more sustainable future will be realized. Considering that Norway scores very high on international indexes on human development and equality as well as human rights, one might think the Sustainable Development Goals do not have much relevance for Norway. As the following account will show, however, this is not the case (Grønningsæter and Stave 2015).

Like most Western countries, Norway does not have an official definition of poverty. Still, the latest figures from the European Union and Statistics Norway show that 11% of the Norwegian population lives under the EU-defined poverty risk line. This figure has been stable, although the issue has had high political priority for more than a decade. In terms of economic equality, Norway has a relatively low level of difference, but it has increased somewhat recently. Regarding social inclusion, whether and how to include poor and undocumented immigrants also represent a challenge. Norway has a very good public health system, but although incidence of drug abuse is relatively low, it scores high on deaths due to drug overdose, and more also needs to be done to reduce communicable diseases like HIV/AIDS. Norway also scores high on various international gender-equality indexes, but a recent parliamentary report concluded that much remains to be done to achieve full equality.

Most observers would claim that in comparative global terms, these are "luxury problems," but for those negatively affected they are real enough. However, there is one issue area where Norway faces real challenges, and it is of utmost importance to achieve a more sustainable planet: the climate change issue. Norway is highly dependent on its fossil-fuel industry: It represented 17% of total GDP in 2014, 330,000 were employed in petroleum-related activities, and total investments in this industry reached an all-time high in 2014. Norway has been a front-runner in providing international assistance to reduce emissions of greenhouse gases internationally, but has not showed much willingness or ability to reduce domestic emissions.

The 2015 Paris Agreement got unprecedented attention in Norway, and many Norwegian environmental groups pointed out the necessity for Norway to reduce its emissions to abide by the agreement. No similar calls regarding climate resulted from the adoption of the Sustainable Development Goals or the other challenges mentioned above. Their adoption got some attention in the Norwegian media, but this was framed largely the same as the Millennium Development Goals, as a need for improved assistance to poor countries. This illustrates the importance of communicating these ambitious and important goals to the outside world so they do not remain within the confines of the UN building. This challenge should be taken up by civil society, UN diplomats as well as academics.

Conclusion

This chapter has reviewed achievements and failures of the Millennium Development Goals and examined the causal link between these goals and actual performance. The formulation of the Millennium Development Goals was largely development-oriented and motivated to increase international aid. Poverty reduction and improvements in nutrition, education, health, and gender equality have been critical issues of concern since the 1940s. In general, the Millennium Development Goals are evaluated positively for their achievements in poverty reduction, gender disparity in school education, gender quality, some of the health-related goals, access to improved drinking water, and mobilization of financial resources for global partnership. On the other hand, the Millennium Development Goals have not succeeded in areas such as decreasing the undernourished population, maternal mortality, attainments of universal access to HIV therapy, sanitation, or environmental sustainability. These underachievements are caused mainly due to the "misfit" of the goals in relation to these problems. In particular, a critical weakness lies in the lack of implementation mechanisms. The one-size-fits-all goals lacked concrete plans for implementation from global to national and local levels, and therefore achievements vary significantly among countries. Results-oriented management goals also failed to include critical issues such as human rights and equality.

Therefore, many have been arguing that the post-2015 development agenda should: (i) set global benchmarks as well as bottom-up goals in line with national circumstances that are practical and clear, (ii) set universal goals for both developing and developed countries, including issues such as climate change, human rights, human security, and governance, along with strengthening cooperation among stakeholders (Koehler, Des Gasper, and Simane 2012; Moss 2010; Poku and Whitman 2011; Vandermoortele 2011). When it comes to the Sustainable Development Goals, goals are required to reflect sustainability challenges, taking into account economic, social, and ecological domains as well as addressing the underachievements of the Millennium Development Goals. Lessons from the Millennium Development Goals tell us that in order for the Sustainable Development Goals to be more effective and "fit" for the purpose, they need to take into account a multilayered approach in which targets are framed in global terms but should be tailored at regional, national, or even organizational levels to provide a menu of options for actors to select those best suited for them (Young et al. 2014).

Regarding health-related lessons, we have focused on Goal 4 regarding child mortality, and particularly emphasized GAVI and Norway in this context, as examples of the major role that partnerships and individual countries can play. The establishment of GAVI cannot be causally linked to the Millennium Development Goals, but those goals have contributed to mobilizing efforts by GAVI (and others) to save children's lives. Despite some shortcomings, there is no doubt GAVI has contributed strongly to a fairly high score on this dimension. When it comes to Norway, the country would have worked to improve global health irrespective of the Millennium Development Goals. However, there is also a direct causal link between these goals and Norway's increasing global health efforts over time.

One reason for the strong international efforts to mobilize support for this goal is the significance of visualization through media mobilization, substantiated by a scientific journal like *The Lancet* and its editor Richard Horton. The combination of his moral and scientific authority is interesting in a "lesson-learned" perspective. The role of the *Lancet* and the brief case studies of GAVI as well as Norway also underline the significance of *leadership* by *individuals*, often forgotten in sober international relations analysis. It may well be that less would have been achieved along this dimension had it not been for these key individuals. Leadership by example has also been characteristic of the role played by Norway as champion for the health-related goals. Although leadership is important, however, these lessons also illustrate the significance of mobilizing *money* from more unconventional sources like the Gates Foundation. Not only may this increase goal achievements in its own right, but it may also forge coordination and cooperation by traditional adversaries. This points to a final lesson, the potential virtue of *combining* UN and non-UN efforts. The United Nations is a necessary venue for securing legitimacy, but smaller and more flexible bodies outside it are often necessary to achieve ambitious goals.

Norway played an important role in realizing the SDG goals. However, we have emphasized that the realization of these goals also represent challenges for a country like Norway.

References

Agwu, Fred Aja. 2011. Nigeria's Non-Attainment of the Millennium Development Goals and Its Implication for National Security. *The IUP Journal of International Relation* 5 (4): 7–19.

Alston, Philip. 2005. Ships Passing in the Night: The Current State of the Human Rights and Development Debate Seen through the Lens of the Millennium Development Goals. *Human Rights Quarterly* 27 (3): 755–829.

Attaran, Amir. 2005. An Immeasurable Crisis? A Criticism of the Millennium Development Goals and why they cannot be measured. *PLoS Medicine* 2 (10): 955–961.

Clemens, Michael, Charles Kenny, and Todd Moss. 2007. The Trouble with the MDGs: Confronting Expectations of Aid and Development Success. *World Development* 35 (5): 735–751.

Development Assistance Committee. 1996. *Shaping the 21st Century: The Contribution of Development Cooperation.* Paris: Organisation for Economic Co-operation and Development.

Easterly, William. 2009. How the Millennium Development Goals Are Unfair to Africa. *World Development* 37 (1): 26–35.

Fukuda-Parr, Sakiko. 2010. Reducing Inequality: The Missing MDG: A Content Review of PRSPs and Bilateral Donor Policy Statements. *IDS Bulletin* 41 (1): 26–35.

GAVI Board, Global Alliance for Vaccines and Immunization Board. 2010. GAVI second evaluation report. Global Alliance for Vaccines and Immunization.

Grønningsæter, Arne Backer, and Svein Erik Stave. 2015. The Global Goals for Sustainable Development: Challenges and Possible Implications for Norway. Discussion paper, Fafo Research Foundation, 1–32.

Hulme, David. 2007. The Making of the Millennium Human Development Meets Results- Based Management in an Imperfect World. Brooks World Poverty Institute working paper 16, 1–26.

Hulme, David. 2009. A Short History of the World's Biggest Promise. Brooks World Poverty Institute Working Paper 100, 1–55.

Hulme, David, and Sakiko Fukuda-Parr. 2011. International Norm Dynamics and "the End of Poverty": Understanding the Millennium Development Goals (MDGs). *Global Governance* (17): 17–36.

Katsuma, Yasushi. 2008. The Current Status and Issues of the Millennium Development Goals: With Focus on Sub-Saharan Africa. *Asia Pacific Research* (10): 97–107.

Koehler, Gabriele, Richard Jolly Des Gasper, and Mara Simane. 2012. Human Security and the Next Generation of Comprehensive Human Development Goals. *Journal of Human Security Studies* 1 (2): 75–93.

Langford, Malcolm. 2010. A Poverty of Rights: Six Ways to Fix the MDGs. *IDS Bulletin* 41 (1): 83–91.

Lie, Sverre O., Dipali Gulati, Halvor Sommerfeldt, and Johanne Sundby. 2011. Millennium Development Goals for Health: Will We Reach Them by 2015? *Tidsskrift for Den norske Legeforeningen* [Journal of the Norwegian Medical Association] 131: 1904–1906.

Manning, Richard. 2010. The Impact and Design of the MDGs: Some Reflections. *IDS Bulletin* 41 (1): 7–14.

McNeill, Desmond, Steinar Andresen, and Kristin Sandberg. 2013. The Global Politics of Health: Actors and Initiatives. In *Protecting the World's Children Immunization Politics and Practices*, ed. Sidsel Roaldkvam, Desmond McNeill, and Stuart Blume, 59–87. Oxford: Oxford University Press.

McNeill, Desmond, and Kristin Sandberg. 2014. Trust in Global Health Governance: The GAVI Experience. *Global Governance* 20: 325–343.

Møgedal, Sigrun, and Benedicte Alveberg. 2010. Can Foreign Policy Make a Difference to Health? *PloS Medicine* 7 (5): e1000274.

Moss, Todd. 2010. What Next for the Millennium Development Goals? *Global Policy* 1 (2): 218–220.

Muraskin, William. 2005. *Crusade to Immunize the World's Children*. Los Angeles: Global BioBusiness Books.

Nelson, Paul J. 2007. Human Rights, the Millennium Development Goals, and the Future of Development Cooperation. *World Development* 35 (12): 2041–2055.

Norwegian Ministry of Foreign Affairs. 2011. Global Health in Foreign and Development Policy [in Norwegian]. Government white paper, 11.

Peterson, Stephen. 2010. Rethinking the Millennium Development Goals for Africa. Harvard Kennedy School Faculty Research Working Paper Series RWP10–046, 1–49.

Poku, Nana K., and Jim Whitman. 2011. The Millennium Development Goals and Development after 2015. *Third World Quarterly* 32 (1): 181–198.

Saith, Ashwani. 2006. From Universal Values to Millennium Development Goals: Lost in Translation. *Development and Change* 37 (6): 1167–1199.

Sandberg, Kristin, and Steinar Andresen. 2010. From Development Aid to Foreign Policy: Global Immunization Efforts as Turning Point for Norwegian Engagement in Global Health. *Forum for Development Studies* 37 (3): 301–325.

Shepherd, Andrew. 2008. Achieving the MDGs: The Fundamentals. *ODI Briefing Paper* 43: 1–4.

UN, United Nations. 2000. We the Peoples: The Role of the United Nations in the 21st Century. New York: United Nations.

UN, United Nations. 2015. The Millennium Development Goals Report 2015. New York: United Nations.

UN, United Nations. n.d. Millennium Development Goals. Available at: http://www.un.org/millenniumgoals/. Accessed August 3, 2015.

Underdal, Arild. 1994. Leadership Theory: Rediscovering the Arts and Management. In *International Multilateral Negotiation: Approaches to the Management of Complexity*, ed. I. William Zartman, 178–197. San Francisco, CA: Jossey-Bass.

UNDP, United Nations Development Programme. 1997. *Human Development Report*. New York: Oxford University Press.

UNGA, United Nations General Assembly. 1948. The Universal Declaration of Human Rights. UN Doc. 217 A (III).

UNGA, United Nations General Assembly. 1970. International Development Strategy for the Second United Nations Development Decade. UN Doc. A/RES/25/2626.

Vandermoortele, Jan. 2011. If Not the Millennium Development Goals, Then What? *Third World Quarterly* 32 (1): 9–25.

Vandemoortele, Jan, and Enrique Delamonica. 2010. Taking the MDGs Beyond 2015: Hasten Slowly. *IDS Bulletin* 41 (1): 60–69.

Young, Oran R. 1991. Political Leadership and Regime Formation. On the Development of Institutions in International Society. *International Organization* 45 (3): 281–309.

Young, Oran R. 2002. *The Institutional Dimensions of Environmental Change: Fit, Interplay, and Scale*. Cambridge, MA: MIT Press.

Young, Oran R. 2008. Institutions and Environmental Change: The Scientific Legacy of a Decade of IDGEC Research. In *Institutions and Environmental Change: Principal Findings, Applications, and Research Frontiers*, ed. Oran R. Young, Leslie A. King and Heike Schroeder, 3–45. Cambridge, MA: MIT Press.

Young, Oran R., Arild Underdal, Norichika Kanie, Steinar Andresen, Steven Bernstein, Frank Biermann, Joyeeta Gupta, et al. 2014. Earth System Challenges and a Multi-layered Approach for the Sustainable Development Goals. Post-2015 Policy Brief 1, 1–4. United Nations University, Institute for the Advanced Study of Sustainability.

8 Corporate Water Stewardship: Lessons for Goal-based Hybrid Governance

Takahiro Yamada

Corporations are increasingly becoming an important element in global environmentalism. They are not only publicizing their own eco-friendly business operations and products, but they are beginning to hold themselves more accountable for the actions of their suppliers as well. Moreover, they are increasingly partnering with international institutions, governments, and civil society organizations alike to address goals of global sustainability.

Nowhere is this effort more pronounced than in the area of water. According to a recent scientific analysis, global fresh water use is one of the three critical earth system processes that are rapidly approaching planetary boundaries (Rockström et al. 2009). In fact, it may not be long before we cross the planetary boundary for water, as the demand for freshwater is currently projected to exceed its supply by 40% in 2030 (2030 Water Resources Group 2012, 5). While the water scarcity problem is essentially a local problem (Whiteman, Walker, and Perego 2012, 314), its cumulative effect can be felt globally even in water-abundant countries. Furthermore, because water is so essential for life, water resource management also has a moral aspect. Universal access to safe drinking water has therefore been on the UN agenda ever since the UN Water Conference in Mar del Plata, Argentina in 1977. However, after more than two decades of endeavor, it became clear that the goal of universal access to safe drinking water was as remote as ever. This led to the inclusion of water access as one of the targets for Millennium Development Goal 7. As will be discussed below, corporations became involved in the process of realizing this target strictly on their own volition. It is this target for water that initially interested them, not other targets or goals. The issue of water therefore provides an excellent case study to understand the role of corporations in global environmental governance.

Not only has the environmental behavior of corporations long remained understudied (Whiteman, Walker, and Perego 2012, 309), but the political behavior of corporations in environmental governance has only recently begun to receive an increasing amount of scholastic attention (Kurland and Zell 2010, 316). To the extent that corporate environmentalism has been studied, however, it has been approached mostly from the perspective of international environmental regimes, such as ozone and climate change regimes (Tienhaara, Orsini, and Falkner 2012), or from the perspective of transnational private governance in such areas as sustainable forestry and climate change (Pattberg 2007; Green 2014). Not as much attention has so far been given to the interaction between intergovernmental organizations and corporations in promoting sustainability in those situations, which are characterized by the absence of regimes, either public or private. The aim of this chapter is therefore to fill in this gap by analyzing how the United Nations has collaborated with businesses to address Millennium Development Goal 7, featuring the UN Global Compact's CEO Water Mandate (hereafter the Mandate). As its name implies, this initiative was designed to bring water issues to the attention of the chief executive officers of water-using companies. It was launched in 2007 at the Global Compact Leaders Summit and has served as the only UN-sponsored multi-stakeholder forum aimed at globally addressing the problem of water scarcity. By examining how the United Nations has used it to elicit cooperation from nonstate actors to attain Millennium Development Goal 7, we should be able to draw implications for the conditions for successful goal-based governance.

With this in mind, I proceed as follows. First, I ask why it is necessary to look at the experience of Goal 7, and then provide a typology of global governance. Second, I briefly review the historical background that has led to the genesis of the Mandate, and then analyze the role of the UN Global Compact vis-à-vis nonstate actors. Lastly, I attempt to draw policy implications from this empirical case study with respect to goal-based governance.

Goal setting as a governance strategy is increasingly becoming the mainstay of today's global governance. In coping with the challenges of climate change, for instance, the international community has set a goal of limiting the average global temperature rise to less than 2° C, and has recently required both developing and developed countries to submit intended nationally determined contributions to meet this goal. This has been touted as a balanced "soft diplomacy approach" by some observers of this process. Will such an approach make governance more effective as well as manageable? Or will it simply make governance more challenging? What kind of effect will such an approach have on global governance, which is increasingly becoming "polycentric" (Abbott and Snidal 2009, 501–545; Ruggie 2013, 78)?

Naturally, different yardsticks can be used to evaluate governance. Effectiveness, efficiency, equity, manageability, and legitimacy are some of the most frequently used gauges of governance. For goal-based governance, similar criteria can be used as well. Oran Young, for instance, argues that the number of goals should be limited and that goals should be clearly defined to provide useful guidance to actors regarding effectiveness (Young, this volume, chapter 2). Similarly, Underdal and Kim stress the importance of having a small set of hierarchically ordered goals (Underdal and Kim, this volume, chapter 10). At the time of this writing, however, the outcome of the UN's Open Working Group, which is the main venue for the negotiation of the Post-2015 Development Agenda, does not meet these requirements because the hard reality facing the Open Working Group does not allow the negotiators to settle for fewer than 17 Sustainable Development Goals and 169 targets. In the words of Ambassador Csaba Kőrösi, co-chair of this working group, this expansive list of goals is "a compromise between what is scientifically advisable, and what is politically feasible" (Kőrösi 2014). This outcome is thus not at all surprising, given the diversity of national priorities, circumstances, and capabilities.

What are the implications of this outcome? It is true that many of these goals are, scientifically speaking, crosscutting in nature—for instance, ensuring sustainable management of water has consequences for other goals, such as food security (Goal 2), energy (Goal 7), and climate change (Goal 13), and their relationships are not necessarily mutually complementary. Yet so far, no agreement has been reached at the global level as to how we should integrate them into a coherent whole. Consequently, such inter-goal integration will likely be deferred to the political judgment of each country. Does this mean that the Open Working Group's effort to set goals and targets has been an outright failure? The answer is obviously no, because each goal is so designed that economic, social, and environmental aspects of sustainable development will be integrated *within* it. As Ambassador Kőrösi (2014) put it, "each SDG [Sustainable Development Goal] has its own unique genetic code for global sustainable development."

While the High-level Political Forum may eventually come to play an important role in ensuring coherence among relevant institutions in and out of the United Nations (Bernstein, this volume, chapter 9), given the political reality discussed above, it is important to reflect on our past experience of goal-based governance to examine how the integration of the three aspects of sustainable development was promoted within the bounds of a single goal. In this respect, Target 7.C of Millennium Development Goal 7 makes an ideal candidate for investigation because it aimed to ensure the availability of safe drinking water and basic sanitation, much like Goal 6, a new SDG for clean water and sanitation. Although the new Goal is more explicit about environmental aspects such as improving water quality, increasing water-use efficiency, and protecting water-related ecosystems, Target 7.C also included concerns for environmental aspects at least implicitly, as connoted by the use of the word "sustainable" (it reads: "Halve, by 2015, the proportion of the population without sustainable access to safe drinking water and basic sanitation.")

Four Modes of Global Governance

Given this nature of the water goal, then, what type of global governance should be adopted? First of all, it goes without saying that we need to realize any problem solving in the international political system has to be performed within the structural constraints of international anarchy. That is, given the absence of a central government, any global governance should necessarily be fragmented. In an anarchic system, the authority enjoyed by international institutions has to be shared with sovereign states. To the extent that international institutions engage in global governance, therefore, they do so primarily through states. This makes global governance necessarily *indirect.* Global governance will become more *indirect,* if the problem to be solved is local in nature, such as a water issue, because its solution depends almost entirely on the willingness and capacity of national and local governments. In this respect, the proposal by Gupta and Nilsson (this volume, chapter 12) for coordinated multilevel responses is a reasonable one.

Within this broad structural constraint, however, more subtle nuances of global governance can be captured by a typology that focuses on two prominent aspects of governance, namely *coerciveness* and *directness,* the elements featured in the "new governance" paradigm adopted in the public administration literature (Salamon 2002, 24–32). *Coerciveness* indicates the extent to which a policy instrument used in global governance restricts the freedom of targeted individual actors, while *directness* indicates the extent to which the international organization authorizing the provision of public goods is involved in the provision of the goods itself. This will give us a two-by-two matrix that classifies global governance into four modes, namely *rule-based public governance, rule-based hybrid governance,*

Table 8.1
Types of Global Governance

	More coercive	Less coercive
More direct	Rule-based public governance Kyoto Protocol *Effective, but less efficient*	Goal-based public governance Post-Kyoto Framework *Less effective, but manageable*
Less direct	Rule-based hybrid governance ILO standards *Effective, but less manageable*	Goal-based hybrid governance Millennium Development Goal 7, Sustainable Development Goal 6 *Efficient, but less effective and manageable (goal displacement)*

goal-based public governance, and *goal-based hybrid governance*, as shown in table 8.1.

Rule-based public governance is the most familiar mode, in which there is inter-governmental concurrence on a common goal as well as on a set of international rules to attain the goal. In this mode, international rules restrict the freedom of governments and other actors in order to ensure their compliance with the rules. Consequently, global governance will be as *coercive* as it can be within the constraints of the international system, and governments will be simultaneously more *directly* involved in regulating the behavior of subnational actors. This mode of governance may be what Young has in mind when he discusses the importance of integrating goals and rules. Since international regimes have both of these elements, this mode of global governance is coterminous with governance based on international regimes (Young and Levy 1999, 14; Young, this volume, chapter 2). The Kyoto Protocol can be listed as an example of this governance mode because it imposes some cost on those who fail to comply with the protocol, and in addition governments are expected to directly control the behavior of subnational actors.

What are the strengths and weaknesses of this governance mode? With all other things being equal, the more *coercive* the instrument of governance, the more *effective* the governance will be. Needless to say, effectiveness is a multifaceted concept; it can mean compliance with rules, changes in the behavior of targeted actors, or the solution of a given problem (Haas, Keohane, and Levy 1993; Young 1999). Here effectiveness simply means changes in the behavior of key actors, which will contribute to the solution of the problem for which a governance system is created. It is also true that the causality between coerciveness and effectiveness has not been fully established as a general law. However, one can plausibly argue that the creation of a legal structure at the domestic level in conformity with the provisions of an international regime is more likely to induce changes in the behavior of key actors because the effective enforcement of national law can ensure their compliance with the regime. However, precisely because governance in this mode entails such a legal structure, the downside of this governance mode will include higher administrative costs, the enlargement of the public sector, and a loss of political support from those whose freedom will be constrained. Consequently, as Conca argues, this mode of governance may not be suitable for problems with local effects like water issues because not all governments have the authority or capacity to control the behavior of subnational actors at the community level (Conca 2006, 49).

Goal-based hybrid governance is the diametrical opposite of rule-based public governance because low levels of *coerciveness* and *directness* characterize this governance mode. As such, neither does goal-based hybrid governance impose any costs or penalties on actors in the private sector, nor does it impose excessive administrative costs on governments. The absence of coerciveness will therefore make this mode an efficient approach. For this reason, this mode would likely be welcomed by both the private sector and governments. Yet there is a drawback to this approach. Because this mode is highly *indirect* in that governance requires the collaboration of private actors, there is always a chance that the private actors will not be as committed to the attainment of the goal as will be its principal, namely the international organization. Especially when the interests of these actors greatly diverge from those of the international organization, it will likely increase the "risk of goal displacement" (Salamon 2002, 31). As will be discussed below, since the solution of many water issues requires a high level of collaboration from the private sector, this mode of governance will most likely be adopted for water governance.

Rule-based hybrid governance and *goal-based public governance* are intermediate modes between rule-based public governance and goal-based hybrid governance. A relatively high level of *coerciveness* and a low level of *directness* characterize rule-based hybrid governance. As such, while the former feature is expected to ensure a certain level of effectiveness, the latter feature reduces the manageability of governance. That is, in rule-based hybrid governance, neither the international organization nor national governments monitor or certify the private actors' compliance with international rules; such functions will be delegated to the private actors themselves. Consequently, the risk of goal displacement is always present. Global governance regarding labor rights may fall into this category because networks of nongovernmental organizations and corporations are implementing the core standards of the International Labor Organization through various private governance schemes. Goal-based public governance, on the other hand, is characterized by a low level of coerciveness and by a high level of directness. As such, it is expected to be less effective, but its manageability is expected to be higher because governments are more directly involved in the governance process. The global framework that has evolved to replace the Kyoto Protocol seems to fit this category because the implementation of the goal of keeping the temperature rise to a minimum level will be largely left to individual governments, which would then exercise some control over the private sector.

Having laid out four different governance modes, which mode will most likely be adopted with respect to water issues? Two variables may become important here. One is the dependence of governments on the private sector's talents and resources for the definition and solution of problems, and the other is the level of controversy surrounding the issues.

Water problems have both of these features. Because there are so many technical issues involved, ranging from assessment of the supply of water and efficient use and recycling of water to the protection of ecosystems, governments need collaboration from private actors, who have the necessary know-how and skills to solve these issues. This need becomes even greater when governmental capabilities are called into question, as in the case of many developing countries. The dependency on the private sector's expertise and resources, therefore, creates the need for *indirect* governance. Moreover, water problems can also be highly controversial. Who has the right to use a local community's scarce resources and how much they are allowed to use are always politically contested because water has to be shared among industry, agriculture, and people. Under such circumstances, it makes political sense to provide "opportunities to cut affected interests into a 'piece of the action'" (Salamon 2002, 30). Apart from the government's general predisposition toward efficiency, therefore, these two factors, namely the governments' dependence on the private sector's capabilities and the contested nature of the issues involved, will make goal-based hybrid governance the most likely candidate for water governance.

If that is the case, however, we are stuck with the risk of goal displacement because more authority will be devolved to the private sector, over which the international organization has less control. How can an international organization adapt to this daunting administrative challenge? The notion of "enablement" skills discussed in the new governance paradigm literature is helpful here. This paradigm prescribes three analytically different types of skills. The first are *activation skills*. With these skills, the principal can be expected to mobilize the "networks of actors increasingly required to address" the problems (Salamon 2002, 16). That is, the international organization as a principal can create opportunities for non-state actors, be it nongovernmental organizations or corporations, to participate in the problem solving by "encouraging the potential partners to step forward and play their roles." The second set of skills the principal is required to use are *orchestration skills* because once actors are activated into a network, the network needs to be maintained to produce collaborative outcomes in line with the goal. The principal, like a symphony conductor, should convey an interpretation of what the common goal entails, while "remaining within the bounds set by the physical capacities of the instruments" used to attain the goal. If orchestration goes well, therefore, the result will be "a piece of music rather than a cacophony" (Salamon 2002, 17). Moreover, what counts in orchestration is those "intangibles" such as vision, knowledge, and persuasion, not material consequences (Ruggie 1982, 1998, 2004; Finnemore 1993, 1996; Katzenstein 1996; Finnemore and Sikkink 1998). In this respect, orchestration assumes the logic of appropriateness. Although this concept of "orchestration" has already been imported into the literature on international organizations, it needs to be emphasized that its essence lies in its effect of avoiding the risk of goal subversion by third-party partners. With this in mind, in International Relations, "orchestration" is thus defined as a process of supporting and/or steering a network of diverse stakeholders for the pursuit of public policy objectives through the use of "a wide range of *directive* and *facilitative* techniques" (Abbott and Snidal 2009, 521 and 565–577; Abbott et al. 2015).

The third and final set of skills that the principal should use are *modulation skills* because persuasion based on the logic of appropriateness may not always work, at least for some actors (Salamon 2002, 17; Risse, Ropp, and Sikkink 1999, 11). To the extent that persuasion is inadequate, the principal needs to rely on "rewards and penalties" in order to elicit cooperative behavior from ill-socialized network partners. However, because goal-based hybrid governance makes use of less *coercive* instruments, the principal needs to provide more positive incentives than disincentives. If the instruments used become too punitive, it will chase the cost-sensitive private partners out of the network.

With these analytical concepts in place, we shall now look at how the United Nations has responded to the problem of water governance in relation to Target 7.C of Millennium Development Goal 7. But, first, let us briefly review the background that led to the formulation of this target.

Analysis of Millennium Development Goal 7

Historical Background

Long before Millennium Development Goal 7 was set, the UN Water Conference held in 1977 in Mar del Plata, Argentina, had set a similar goal of universal access to safe drinking water and sanitation services. Subsequently, in accordance with this goal, the UN General Assembly declared the 1980s the International Drinking Water Supplies and Sanitation Decade. Yet, by the end of that decade, it became clear that little progress had been made in meeting the goal. As a result, water experts began to call for a more comprehensive approach, which became known as integrated water resource management. The "Dublin Principles," ratified at an international conference on water just preceding the UN Conference on Environment and Development, embodied this new comprehensive approach. The basic tone of these principles was, however, extremely neoliberal because these principles treated water essentially as an economic good.[1] As international development agencies began to endorse the idea of integrated water resource management, the privatization of water supplies came to be seen as the panacea for all water problems. Major multinational water suppliers such as Suez Lyonnaise des Eaux and Vivendi jumped on the bandwagon and began supplying water in many countries.

After a while, this neoliberal ideology began to draw criticism from water activists because no improvements were yet to be made in providing universal access to water and sanitation. In certain water-stressed regions, the situation even worsened, giving rise to tense anti-privatization movements. For instance, in South Africa, a violent protest broke out because private water companies cut off water supplies (Conca 2006, 238–239), and in India, a Coca-Cola plant in Plachimada, Kerala, was forced to shut down because the company was believed to be depleting the community's ground water (Bywater 2012, 208–209). Before too long, these movements became transnationally linked, largely owing to the activities of the World Social Forum, a counterpart to the World Economic Forum (Pigman 2007, 128–129). Through these protests, activists tried to drive home the point that only a well-managed, democratically accountable public sector could ensure universal access to water and sanitation.[2]

By the early 2000s, UN institutions began to respond to this mounting criticism. In 2000, for instance, the UN Millennium Summit endorsed the goal of reducing the proportion of the world's population lacking affordable access to safe water by half by 2015, and this was followed by the decision at the World Summit on Sustainable Development to make water one of its key themes for sustainable development (Bakker 2012, 26). Once again, therefore, a goal and a target were set to remind the international community of the importance of water access.

Activation

Did the United Nations use its activation skills to form a network of both governments and nongovernmental actors? If the United Nations were to perform such a task, it should be the job of the UN Global Compact Office because the UN Global Compact was launched to encourage businesses to commit themselves to the Millennium Development Goals as well as to the UN Global Compact's 10 universal principles in the areas of human rights, labor rights, environment, and corruption.[3] It was indeed as part of this mission that the UN Global Compact Office initiated the Mandate in the summer of 2007. The idea was to create a network of businesses, nongovernmental organizations, and governments, mainly to assist businesses in the development, implementation, and disclosure of water sustainability practices, thereby contributing to the attainment of Target 7.C of Goal 7.

Genesis of the CEO Water Mandate

Let us first look at how this initiative came about. According to Gavin Power, Deputy Director of the UN Global Compact Office, who has served as the Mandate's Secretariat since its inception, it was a handful of high-volume corporate water users that provided the original idea for the Mandate (Power 2014).

This is how the idea was conceived. Since most of the companies committed to the UN Global Compact had only focused on fleshing out the UN Global Compact's 10 principles, businesses were less concerned with addressing the Millennium Development Goals in general. Many of them found it difficult to understand why development issues such as poverty were important for businesses. This led to the UN Global Compact Office's decision to raise awareness with businesses about the Millennium Development Goals. Gradually, some companies began to have internal discussions about how they could expand their corporate social responsibility strategy to incorporate a development component. The UN Global Compact Office also began its discussions regarding how to extend the 10 principles into some concrete issue platforms. At that time, a few companies approached Mr. Power and suggested water as a possible area for a focus, pointing out to him that the water issues were becoming "increasingly important and material" for them (Power 2014). Power then decided that water would be a good choice because it is directly relevant to the UN Global Compact's environmental and human rights principles. Power recalls that "it was really the companies that came forward and said, 'it would be interesting if you would consider launching a Global Compact initiative on water to help us go 'deep' into the issue'" (Power 2014). In response to this corporate initiative, the UN Global Compact Office subsequently arranged an informal meeting with company representatives at the Swedish Embassy in Washington, DC.

The meeting was held in 2006 with representatives from six companies, including Coca-Cola, Levi's, and Nestle. This meeting basically turned into a brainstorming session that shaped the outline of the Mandate. They agreed early on that this should be an initiative for large-scale corporate water users, and not for water utility companies or water distributors. The company representatives argued that through this initiative, they wanted to understand how they could meet challenges where they were beginning to experience water stress. They also emphasized that they were facing reputational and regulatory risks because if they did not act responsibly in water-stressed regions, they could easily be accused of taking water away from the local community and might face costly regulations. This awareness, shared by the participating companies, led to the selection of six focal areas, which were dubbed "the six elements" of the Mandate: direct operations, supply chain and watershed issues, community engagement, collective action, public policy engagement, and transparency (United Nations Global Compact 2011b). They also felt that the best way to elicit cooperation from the private sector was to issue a set of recommended actions, then have the companies report on them. Subsequently, this idea was conveyed to UN Secretary-General Ban Ki-moon, who then turned it into a major deliverable at the 2007 Global Compact Leaders Summit.

Institutionalization of the Mandate

Initially, the idea was simply to ask the endorsing companies to commit themselves to the Mandate's six elements and to have them report on their plans and actions. The UN Global Compact Office, however, subsequently decided to create an institutional apparatus equipped with some organizational resources.

The Mandate now includes a steering committee, annual working conferences, endorser-only meetings, working groups, and a Secretariat. The steering committee is primarily charged with strategic decision making and administrative oversight, and is composed of 10 endorsing company representatives chosen from five different regions, as well as *ex officio* members from the Secretariat. Each stakeholder member serves on the committee for two years. The steering committee has also come to include nonvoting special advisors representing nonbusiness stakeholder interests, such as nongovernmental organizations and governments. The main task of endorser-only meetings is to supplement the steering committee functions, especially when there are issues that a broader constituency should be informed about. Working groups are usually composed of a small number of company and civil society representatives, and meet through conference calls mainly to discuss work plans, drafts of guides, and other outputs.

Lastly, the Mandate has a unique arrangement for its Secretariat because it is based on a partnership with a specialized research institute, not affiliated with the United Nations. Jason Morrison from the Pacific Institute, specializing in environmental protection, economic development, and social equity issues, works in tandem with UN Global Compact Deputy Director Gavin Power to serve as the Mandate's Secretariat. In practice, the Pacific Institute functions as the initiative's "operational arm." It coordinates the working conferences, conducts research on topics relevant to the initiative's work streams, and helps the working groups to develop guides on various aspects of water management. As Power puts it, this arrangement is "a match made in heaven" (Power 2014). None of the UN Global Compact's other "issue platforms" has this type of arrangement. This can be explained by the absence of a UN specialized agency, which is capable of supplying technical advice regarding water. True, there is UN-Water, which serves an interagency organization performing coordination functions within the UN system on issues related to water, but it is not a technical organization capable of supplying expertise on water.

In short, not only has the UN Global Compact mobilized businesses to play an active role in addressing Goal 7's target on water, it has also helped create the network of business and nonbusiness actors concerned with water issues by providing a formalized institutional arena.

Assessment

The cursory overview of the initiative's institutions above reveals that its annual working conferences are precisely where endorsing companies engage in the Mandate's networking activities; they share practices, discuss complex issues, identify practical solutions, and get feedback from other stakeholders. In order to ascertain the intensity of their interaction, we thus need to ask the following questions: How well have the endorsing companies actually attended these meetings, and to what extent has their attendance been matched by the attendance of nonbusiness stakeholders?

Let us first look at the changes in the number of endorsing companies in order to measure the magnitude of overall corporate interest in the Mandate. When the initiative was launched in 2007, there were only five endorsing companies. By mid-2010, the number had reached 75, and currently 129 companies have come to endorse the initiative. Admittedly, it is a very small issue platform, attracting only a fraction of the total number of companies participating in the UN Global Compact, which is estimated to be about 8,000. Even when compared to Caring for Climate, the UN Global Compact's other environmental issue platform, the Mandate's membership is significantly smaller, for Caring for Climate now has over 400 signatories. Nevertheless, the change in the number of the endorsing companies clearly indicates the intensification of corporate interest in water issues.

Moreover, of these endorsing companies, about a third seem to be regularly attending the Mandate's working conferences; for instance, in 2010, 25 out of 75 endorsers attended these conferences, and in 2013, 41 out of 122 endorsers attended them on average. Nongovernmental organizations and government organizations are also participating in these conferences fairly vigorously. Table 8.2 gives the breakdown of participation by these three actors in percentiles for the period of 2008 through 2013; corporate participation has averaged 54.4%, while civil society and government participation averaged 26.1% and 19.5%, respectively. From these data, therefore, one can confirm that the Mandate with its permanent institutional

Table 8.2
Average Participation by Stakeholders (in percent)

	2008	2009	2010	2011	2013	Average
Corporation	59.1	47.7	53.2	58.0	54.1	54.4
Nongovernmental organizations	20.2	28.9	31.9	21.3	28.3	26.1
UN/Government	20.8	23.4	15.0	20.7	17.7	19.5

Sources: Calculated from lists of participants attached to the meeting summary for CEO Water Mandate Working Conference (UN Global Compact), Inaugural, Second, Third, Fourth, Fifth, Sixth, Seventh, Eighth, Eleventh, and Twelfth meetings (the meeting summaries for the Ninth and Tenth CEO Water Mandate Working Conferences are unavailable).

structure in place has served as a vibrant platform for multi-stakeholder networking. One can therefore say that, all in all, the UN Global Compact Office has performed its *activation* function quite effectively. Not only has it urged private actors to step forward to play an important role in implementing the Millennium Development Goals, but it has also created and institutionalized a network of stakeholders to promote dialogue concerning water issues.

Orchestration

Then, the question becomes whether the UN Global Compact Office has also engaged in *orchestration* to create a common vision or knowledge among the stakeholders. If such an orchestration has indeed been performed, we should find out how it has been carried out, and what shared interpretation it has produced regarding the role of business in relation to water issues.

At the Mandate's Second Working Conference, held in Stockholm, August 2008, it was agreed that the biggest challenge for the Mandate was to define the respective roles for public and private actors in providing universal access to safe drinking water and sanitation (CEO Water Mandate 2008, 8). This then led to the creation of two work streams: one on corporate engagement with water policy, and the other on water and human rights (CEO Water Mandate 2009c, 4–7). Of these two, initially the former attracted more attention due to its connection with Millennium Development Goal 7. The summary of the meeting stresses the importance of this work stream as follows: "Globally accepted policy objectives such as the UN MDGs can not only steer national/regional water policies, but also corporate water management practices (and policy engagements activities). ... Companies can *take actions to promote strong public regulatory frameworks and water governance*" (CEO Water Mandate 2009b, 14, emphasis added).

Let us first look at how shared knowledge concerning this theme has emerged. In this connection, it is important to point out that the UN Global Compact Office's orchestration has made full use of the expertise provided by nongovernmental organizations. To be more specific, from the outset, the Pacific Institute and the World Wide Fund for Nature clearly set the tone for the ensuing discussion between endorsers and other stakeholders (CEO Water Mandate 2009a). One can therefore argue that the contribution of nongovernmental organizations to the Mandate's final outputs, namely the Framework for Responsible Business Engagement with Water Policy (hereafter, the Policy Engagement Framework) and the Guide to Responsible Business Engagement with Water Policy, was substantial.[4]

Then, what interpretation have business and nonbusiness participants come to share with respect to the role of business? We can summarize the Mandate's argument as follows. Water is a scarce and nonsubstitutable resource shared by a multitude of users. As such, water availability and its quality are affected by the water use and discharge practices of all the users. Companies, which use water for industrial purposes, therefore need to collaborate with local communities and governments to jointly reduce water risks. To be sure, many progressive companies have already begun to make an effort to understand their water use and discharge within their fence lines as well as in their supply chains. Some companies, such as Coca-Cola, PepsiCo, and Nestlé, for instance, now have a self-professed goal of becoming "water neutral," and thus constantly monitor their impact on water resources (Dauvergne and Lister 2013, 65). There is no doubt this is an important step companies can take in reducing the negative impact of their water use and discharge. Such within-the-fence-line water management, however, is not considered sufficient because one drop of water conserved through one company's internal effort may mean different things for different communities, depending on the amount of water that other users consume, as well as on government's capacity to police illegal water withdrawals and substandard water discharges.

The Mandate's Policy Engagement Framework thus identifies five primary scales for water policy engagement: *internal* (internal efficiency, pollution control); *local* (water supplies and sanitation at the level of municipalities); *regional* (overseeing judicious basin management); *national* (ensuring the enactment of appropriate legislative and institutional arrangements regarding the supply and use of water, equitable access, and the quality of water); and *global* (engaging with development agencies, international financial institutions, and nongovernmental organizations for international advocacy and development of best practices and new standards; CEO Water Mandate 2010a, 3 and 5–6).

More specifically, by engaging with local communities, companies can share what they have learned from water footprint analyses with other users and concerned stakeholders. In light of competing water demands in local communities, companies need to take such steps to reduce "the risk of future water-related disputes or disruptions" (CEO Water Mandate 2009a, 6). Companies also should assist in developing local water systems by providing technologies such as clean water technologies and rainwater harvesting techniques as well as providing funds for managing watersheds.

Only through this type of local engagement can companies help reduce their reputational risks to maintain their social license to operate.

In addition, companies need to engage with governments because "the way in which governments manage water for all users" affects the level of social and environmental risk companies will face (CEO Water Mandate 2009a, 7). When governments do not have effective and fair policies regarding water infrastructure development, water allocation and pricing, management of water supplies, delivery of sanitation services, and protection of natural systems, companies' efforts to improve their own water efficiency will at the end of the day amount to nothing. Companies therefore need to engage in public policy processes to ensure that governments enact appropriate legislative and institutional arrangements regarding the supply and use of water, equitable access to water, and the quality of water, while explicitly articulating their concern for the public interest. Otherwise, companies will face high water risks because a free-for-all situation obtains. In short, both community engagement and responsible corporate engagement in public policy are regarded as an important business contribution to effective water governance.

This shared interpretation is therefore informed by a concern for risks that businesses may face when they fail to engage in water policy and collective action. In particular, *reputational* and *regulatory risks* were discussed during the course of deliberation (CEO Water Mandate 2014, 4–6). Some corporations are high-volume water users and are thus likely to face physical risks in situations of water scarcity. However, physical risks are not the only risks they should be mindful of. They should also pay attention to *reputational risks* because, in the absence of effective water governance, corporations may run the risk of being accused of exploiting the community's water. In the worst-case scenario, they might even lose their *social license* to operate as their reputation becomes irreparably tarnished. They will also likely face *regulatory* risks because, should companies fail to adequately cope with such "reputational risks," governments will likely be pressed to tighten withdrawal permits or revoke their legal license to operate. Companies will also face another type of *regulatory* risk when the government fails to police polluters of ambient water or illegal water users, or fails to allocate water permits fairly according to needs because, under such circumstances, the amount of water that companies can use will inevitably be reduced. Yet, for various reasons, in many developing countries, governments do not generally perform regulatory functions in an adequate fashion.[5]

Given these risks, therefore, it makes sense to businesses to mobilize both governments and communities to establish a well-functioning institutional framework for water governance. Policy engagement and community engagement have thus become top priorities for the endorsing companies of the Mandate. It is interesting to note that while some of the endorsing companies were well aware of these risks right from the start, it was nongovernmental organizations that conceptualized these risks and provided policy prescriptions as to how to mitigate them (CEO Water Mandate 2009b, 7–8 and 14). Indeed, without this epistemic input from civil society organizations, it is hard to imagine that any shared knowledge could ever have emerged. In other words, the orchestration by the UN Global Compact Office was made possible through the orchestrated collaboration with civil society organizations.

More recently, the Mandate has begun to promote this shared understanding under the label of "corporate water stewardship." While prescribing the same responsible behavior as the Policy Engagement Framework and Guide, this concept prompts companies to be more assertive about how they can become the stewards for water. As part of this corporate stewardship, the Mandate issued a communiqué to government leaders who gathered at the 2012 UN Conference on Sustainable Development to "urge them to make water and sanitation a key priority," and has subsequently advocated the inclusion of water and sanitation issues in the 2030 Agenda for Sustainable Development as a stand-alone goal in the UN's Open Working Group process.[6]

Modulation

Will the provision of a common interpretation regarding the role of business be adequate to avoid the risk of goal disruption associated with goal-based hybrid governance? Can companies be trusted to internalize the shared knowledge?

There is obviously a temptation for engaging in "greenwashing" or "blue-washing." That is, companies may claim that they are doing something good for the environment, while they are actually not, or they may claim that they are collaborating with the United Nations without actually contributing to the cause. There is therefore no guarantee that companies act responsibly at all times. Yet deception can prove to be costly for companies, should the truth come out, because it may lead to the withdrawal of their social license, or even short of that, it could lead to a loss of investor or consumer confidence. The market will eventually punish dishonest companies if they are transparent to the market.

The question therefore boils down to how the UN Global Compact Office has actually incentivized companies to be transparent. In this connection, the Mandate's disclosure requirement should be mentioned. As part of their commitment to the Mandate, the endorsing companies are required to annually disclose their strategies and actions related to its six elements in a Communication on Progress. Companies get delisted if they fail to meet this requirement. Some companies, failing to meet this requirement, have already been delisted. Moreover, like other UN Global Compact platforms, companies are organized into three categories, namely "Learner," "Active," and "Advanced," depending on the level of submitted Communications on Progress. This categorization is intended to "encourage and challenge participants to use more sophisticated methodology and release more detailed" Communications on Progress (United Nations Global Compact 2014, 3), on the assumption that the more detailed the companies' disclosed information is, the more confident other stakeholders can feel about these companies. Companies are thus provided with powerful incentives to be responsible because stakeholders such as consumers and investors can penalize irresponsible companies. In short, the linkage between transparency and the market is consciously built into the system to create an incentive structure for companies to be responsible.

Conclusion

In this chapter, I have provided a typology of global governance on the basis of coerciveness and directness and have argued that water issues would fall into *goal-based hybrid governance* characterized by a low level of directness and a low level of coerciveness. I have also pointed out that this mode of governance will suffer from the risk of goal disruption between the principal and its third-party partners. Thus, I have expected the international organization in charge to make use of *activation, orchestration,* and *modulation skills* to reduce this risk.

I have found that the UN Global Compact Office has indeed engaged in *activation.* It has not only mobilized businesses into a network of stakeholders to address Target 7.C of Goal 7, but it has also institutionalized the network. This in turn has created an enabling environment for the UN Global Compact Office's subsequent *orchestration,* whereby the UN Global Compact Office has provided an interpretation of what Target 7.C means for business. Thereafter, this interpretation has come to be shared by the Mandate participants and has led to the conceptualization of corporate water stewardship. In addition, the UN Global Compact Office has also performed its *modulation* function by creating an incentive structure for companies to commit themselves to the requirements of corporate water stewardship.

The implications for goal-based governance are therefore as follows. First, setting a goal and a target such as Goal 7 and Target 7.C has made a difference in the way the United Nations has approached water issues. It has definitely facilitated the mobilization of the private sector. Indeed, without the goal and the target, it might have been difficult, if not impossible, for the UN Global Compact Office to engage in goal-based hybrid governance in the first place. Having said that, however, one cannot assume that *activation* will always be followed by *orchestration*. In the case of the Mandate, it so happened that the UN Global Compact Office has performed orchestration as well, but whether or not the UN Global Compact Office always engages in orchestration is open to question. Needless to say, had it not performed the orchestration function, no social construction of business' role with respect to water governance would have been possible.

Yet, the question still remains whether the UN Global Compact Office's effort to engage in goal-based hybrid governance has made a difference in terms of attaining Goal 7 itself. True, as far as Target 7.C is concerned, the world has met the target with respect to access to improved sources of water five years ahead of schedule. However, how much of this is attributable to corporations' behavioral change is unknown. Neither do we know the extent to which corporations have engaged in water policy as a result of the UN Global Compact Office's orchestration and modulation. Interestingly, however, a survey conducted by the Global Compact Office in 2009 shows that of all the partnerships the survey's respondents had been involved in, 85% were actually aimed at the implementation of Goal 7, and out of these partnerships, about 70% had advocacy as a dominant objective, while 44% listed governments as partners (United Nations Global Compact 2011a, 5–57). Yet it is impossible to tell how much of this is attributable to the outcome of the Mandate itself.

Nevertheless, the record shows that the endorsing companies of the Mandate have consistently called on national governments to prioritize water issues, and also to incorporate water, sanitation, and hygiene into the Post-2015 Development Agenda. More importantly, some of the endorsing companies have actually begun to engage in water policy reforms in several water-stressed countries in collaboration with the International Finance Corporation.[7] Such corporate stewardship is extremely important in light of the fact that national governments in water-stress areas may not always have an incentive to address the ecological challenges of water scarcity. It is therefore not an exaggeration to say that the success of goal-based governance for water will largely depend on the level of corporate engagement with governments. It is in this respect that the UN Global Compact Office's "new governance" has had an important effect on how different stakeholders, be it state or nonstate actors, come to align their behavior with a common global goal to enhance the effectiveness of polycentric governance for water.

Notes

1. This ideological stance is known as "market environmentalism," "liberal environmentalism," or "green neoliberalism" (Bakker 2004, 2010, and 2012; Bernstein 2001; Goldman 2005).

2. They insisted on the continued monopoly of the state authority, the protection of public-sector jobs, and the continuation of subsidized water services (Conca 2006, 247; Bakker 2012, 21).

3. The launching of the UN Global Compact was proposed by UN Secretary-General Kofi Annan at the 1999 World Economic Forum in order to "give a human face to the global market" (Gregoratti 2012, 96).

4. The framework was delivered at the UN Global Compact Leaders Summit in June 2010, and the full-length guide was released at the Sixth Working Conference held in Cape Town, South Africa, in November 2010 (CEO Water Mandate 2010b, 6).

5. The obstacles include corruption, a lack of awareness, competing demands for water, and a lack of administrative resources (CEO Water Mandate 2014, 9).

6. A CEO from Unilever, one of the Mandate's active endorsing companies, called for long-term water planning, improved efficiency particularly in agriculture, increased investment in infrastructure for efficient water and sanitation services delivery, and fair allocation of water among different users (United Nations Global Compact 2014, 2–3).

7. Since 2010, the 2030 Water Resources Group managed by the International Finance Corporation has led multi-stakeholder pilot country programs for water policy reforms in several countries (2030 Water Resources Group 2009, I).

References

2030 Water Resources Group. 2009. Charting Our Water Future: Economic Frameworks to Inform Decision-Making.

2030 Water Resources Group. 2012. The Water Resources Group: Background, Impact and the Way Forward. Briefing report prepared for the World Economic Forum Annual Meeting in Davos-Klosters, Switzerland, January 2012.

Abbott, Kenneth W., Philipp Genschel, Duncan Snidal, and Bernhard Zangl. 2015. *International Organizations as Orchestrators*. Cambridge, UK: Cambridge University Press.

Abbott, Kenneth W., and Duncan Snidal. 2009. Strengthening International Regulation Through Transnational New Governance: Overcoming the Orchestration Deficit. *Vanderbilt Journal of Transnational Law* 42: 501–578.

Bakker, Karen. 2004. *An Uncooperative Commodity: Privatizing Water in England and Wales*. Oxford: Oxford University Press.

Bakker, Karen. 2010. *Privatizing Water: Governance Failure and the World's Urban Water Crisis*. Ithaca: Cornell University Press.

Bakker, Karen. 2012. Commons Versus Commodities: Debating the Human Right to Water. In *The Right to Water: Politics, Governance and Social Struggles*, ed. Farhana Sultana and Alex Loftus, 19–44. London, New York: Earthscan.

Bernstein, Steven F. 2001. *The Compromise of Liberal Environmentalism*. New York: Columbia University Press.

Bywater, Krista. 2012. Anti-Privatization Struggles and the Right to Water in India: Engendering Cultures of Opposition. In *The Right to Water: Politics, Governance and Social Struggles*, ed. Farhana Sultana and Alex Loftus, 206–222. London: Earthscan.

CEO Water Mandate. 2008. CEO Water Mandate Second Working Conference, August 21–22, 2008, World Water Week, Stockholm, meeting summary.

CEO Water Mandate. 2009a. From Footprint to Public Policy: The Business Future for Addressing Water Issues. Discussion Paper.

CEO Water Mandate. 2009b. CEO Water Mandate Third Working Conference, March 15–17, 2009, World Water Forum, Istanbul, meeting summary.

CEO Water Mandate. 2009c. CEO Water Mandate Fourth Working Conference, August 16–18, 2009, World Water Week, Stockholm, meeting summary.

CEO Water Mandate. 2010a. Framework for Responsible Business Engagement with Water Policy.

CEO Water Mandate. 2010b. CEO Water Mandate Sixth Working Conference, November 14–17, 2010, Cape Town, South Africa, meeting summary.

CEO Water Mandate. 2014. Shared Water Challenges and Interests: The Case for Private Sector Engagement in Water Policy and Management. Discussion paper.

Conca, Ken. 2006. *Governing Water: Contentious Transnational Politics and Global Institutional Building*. Cambridge, MA: MIT Press.

Dauvergne, Peter, and Jane Lister. 2013. *Eco-Business: A Big-Brand Takeover of Sustainability*. Cambridge, MA: MIT Press.

Finnemore, Martha. 1993. International Organizations as Teachers of Norms: The United Nations Educational, Scientific, and Cultural Organization and Science Policy. *International Organization* 47 (4): 565–597.

Finnemore, Martha. 1996. Norms, Culture, and World Politics: Insights from Sociology's Institutionalism. *International Organization* 50 (2): 325–347.

Finnemore, Martha, and Kathryn Sikkink. 1998. International Norm Dynamics and Political Change. *International Organization* 52 (4): 887–917.

Goldman, Michael. 2005. *Imperial Nature: The World Bank and the Making of Green Neoliberalism*. New Haven: Yale University Press.

Green, Jessica. 2014. *Rethinking Private Authority: Agents and Entrepreneurs in Global Environmental Governance*. Princeton, Oxford: Princeton University Press.

Gregoratti, Catia. 2012. The United Nations Global Compact and Development. In *Business Regulation and Non-State Actors: Whose Standards? Whose Development?* ed. Darryl Reed, Peter Utting, and Ananya Mukherjee-Reed, 95–108. London: Routledge.

Haas, Peter M., Robert O. Keohane, and Marc A. Levy, eds. 1993. *Institutions for the Earth: Sources of Effective Environmental Protection*. Cambridge, MA: MIT Press.

Katzenstein, Peter J. 1996. *The Culture of National Security: Norms and Identity in World Politics*. New York: Columbia University Press.

Kőrösi, Csaba. 2014. Keynote speech at the Science and the Sustainable Development Goals Conference sponsored by the United Nations University's Institute for the Advanced Study of Sustainability, Tokyo Institute of Technology, and POST 2015 Project of the Ministry of Environment, Tokyo, Japan, November 15, 2014.

Kurland, Nancy B., and Deone Zell. 2010. Water and Business: A Taxonomy and Review of the Research. *Organization & Environment* 23: 316–353.

Pattberg, Philipp H. 2007. *Private Institutions and Global Governance: The New Politics of Environmental Sustainability*. Northampton, MA: Edward Elgar.

Pigman, Geoffrey A. 2007. *The World Economic Forum: A Multi-stakeholder Approach to Global Governance*. New York: Routledge.

Power, Gavin. 2014: Interview by author with Mr. Power, Deputy Director, UN Global Compact Office. New York City, 28 February.

Risse, Thomas, Stephen C. Ropp, and Kathryn Sikkink, eds. 1999. *The Power of Human Rights: International Norms and Domestic Change*. Cambridge, UK: Cambridge University Press.

Rockström, Johan, Will Steffen, Kevin Noon, Åsa Persson, F. Stuart Chapin III, Eric F. Lambin, Timothy M. Lenton, et al. 2009. A Safe Operating Space for Humanity. *Nature* 461: 472–475.

Ruggie, John G. 1982. International Regimes, Transactions, and Change: Embedded Liberalism in the Postwar Economic Order. *International Organization* 36 (2): 379–415.

Ruggie, John G. 1998. *Constructing the World Polity: Essays on International Institutionalization*. London: Routledge.

Ruggie, John G. 2004. Reconstituting the Global Public Domain: Issues, Actors, and Practices. *European Journal of International Relations* 10 (4): 499–531.

Ruggie, John G. 2013. *Just Business: Multinational Corporations and Human Rights*. New York: W. W. Norton.

Salamon, Lester M, ed. 2002. *The Tools of Government: A Guide to the New Governance*. New York: Oxford University Press.

Tienhaara, Kyla, Amandine Orsini, and Robert Falkner. 2012. Global Corporations. In *Global Environmental Governance Reconsidered*, ed. Frank Biermann and Philipp Pattberg, 45–67. Cambridge, MA: MIT Press.

United Nations Global Compact. 2011a. United Nations Global Compact Annual Review 2010—Anniversary Edition. New York: United Nations Global Compact Office.

United Nations Global Compact. 2011b. The CEO Water Mandate: An Initiative by Business Leaders in Partnership with the International Community. New York: United Nations Global Compact Office.

United Nations Global Compact. 2014. Post-2015 Agenda and Related Sustainable Development Goals, Issue Focus: Water and Sanitation and the Role of Business. UN Global Compact Briefing Series, Issue Paper 6.

Whiteman, Gail, Brian Walker, and Paolo Perego. 2012. Planetary Boundaries: Ecological Foundations for Corporate Sustainability. *Journal of Management Studies* 50 (2): 307–336.

Young, Oran R. 1999. *Governance in World Affairs*. Ithaca, London: Cornell University Press.

Young, Oran R., and Marc A. Levy. 1999. The Effectiveness of International Environmental Regimes. In *The Effectiveness of International Environmental Regimes: Causal Connections and Behavioral Mechanisms*, ed. Oran R. Young, 1–32. Cambridge, MA: MIT Press.

III Operational Challenges

9 The United Nations and the Governance of Sustainable Development Goals

Steven Bernstein

The purpose of the Sustainable Development Goals is to mobilize action to address systemic challenges across economic, social, and ecological dimensions of sustainable development. However, even if the goals were perfectly designed according to criteria identified by other contributors to this volume—fully coherent, built around consensual knowledge, action-oriented, with multilayered differential targets, and adapted to national capacities and circumstances—they would still require appropriate governance arrangements to diffuse them and integrate them into institutions, policies, and practices. The United Nations has an important leadership role to play in such governance efforts, even as the nature of goals means that governance cannot rely solely on traditional tools of multilateralism. The challenge, then, is how to balance requisite political leadership, political authority, and steering at the global level with the reality that action and resources must be mobilized also at regional, national, and local levels, and by a wide range of public and private actors, partnerships, and networks. Put another way, the Sustainable Development Goals, ideally, can provide the direction of steering—the normative underpinning, compass, and guideposts—for the pursuit of sustainable development, but do not, simply by virtue of their articulation, provide the authority, tools, or means required.

This chapter focuses especially on the United Nations' leadership role in governance of the Sustainable Development Goals for two simple reasons: governments set the Sustainable Development Goals through the United Nations, and they have mandated the United Nations to follow up, monitor, and review all commitments related to sustainable development, as well as to mobilize means of implementation. The 2012 UN Conference on Sustainable Development created the new High-level Political Forum on Sustainable Development (hereafter, "High-level Political Forum") to lead this effort. Governments gave it an appropriately ambitious mandate for this purpose, but limited direct authority or resources of its own to carry it out. Under such circumstances, "orchestration" is an apt metaphor for the leadership the High-level Political Forum can provide: it can serve as a directorial platform through which governments can promote the coordination and combination of policies to diffuse and integrate the Sustainable Development Goals into global, country-level, and marketplace policies and practices (Abbott and Bernstein 2015).

The mandated functions of the High-level Political Forum provide a useful lens through which to analyze the prospects, challenges, and appropriate role of the UN system in mobilizing the commitments, resources, and mechanisms to produce action and follow up on the Sustainable Development Goals. Yet, as will become clear in the pages that follow, the High-level Political Forum, while politically important and central, is but a small node in the overall governance arrangements, and the mobilization that can flow from them, to catalyze the resource and actions for goal attainment. Thus, while agreeing with Underdal and Kim's (this volume, chapter 10) admonishment that "goal attainment" is a different and "far more demanding" task than goal setting, this chapter argues that the UN system—and the High-level Political Forum as a possible "orchestrator of orchestrators" within the crowded field of relevant governance institutions and arrangements—is a central, and necessary, node in creating, supporting, and enabling the governance environment to encourage goal attainment.

To assess the conditions and prospects of UN-led governance of the Sustainable Development Goals, I first explore in more detail the nature of the governance challenge. I then outline the core purposes required for governance of the Sustainable Development Goals—coherence, orchestration, and legitimacy—and examine the conditions under which the United Nations can meet those requirements. I also offer a preliminary assessment of prospects and challenges in meeting those conditions in light of the means mandated to it and evolving governance arrangements within, and beyond, the UN system.

The Governance Challenge of Sustainable Development Goals

Governance of the Sustainable Development Goals faces three challenges that stem from their very constitution: their nature as goals; the normative character of sustainable development that demands integrative and coherent governance; and their scope in relation to the wider 2030 Agenda (UNGA 2015) and social, economic, and environmental governance more generally. In examining these challenges, minimum requirements for governance can be identified.

Goals versus Rules

Although goals have long been a feature of governance in many settings, less is known about the conditions for effective governance via "goals" as compared to rules. Whereas Young (this volume, chapter 2) and Underdal and Kim (this volume, chapter 10) provide a variety of plausible hypotheses on conditions for effective goal setting and goal attainment, none have been tested. More fundamentally, despite a great deal of work on measuring, monitoring, and assessing progress on the Millennium Development Goals, even less is known about the causal relationship of goals to outcomes.[1] The wide range of drivers of sustainable development must be considered in any analysis of the role of goals and governance mechanisms in changing behavior and outcomes. Thus, an important element of any governance arrangement must be mechanisms of learning and analysis to improve our knowledge of drivers of sustainable (and unsustainable) development and the relationship between goals, policies, and plans, enabling mechanisms such as capacity building and learning, and outcomes. Many assessments of the achievements of the Millennium Development Goals can point to outputs, such as participation of stakeholders, or changing the priorities of particular policies, such as poverty eradication in developing countries. However, how these outputs contributed to goal outcomes, or how the policies and practices specifically motivated by goals directly contributed to outcomes, as compared to, or in combination with, factors such as economic liberalization or other drivers of development over the last 15 years, has not been studied (an exception is Andresen and Iguchi, this volume, chapter 7, which focuses specifically on Millennium Development Goal 4 and its child mortality target).

In addition, as other contributors to this volume have highlighted (for example, Haas and Stevens, chapter 6; Underdal and Kim, chapter 10), the experience of the Millennium Development Goals has taught us that goals may interact in unintended ways. For example, efforts to reduce poverty and hunger through modern agriculture can hinder environmental sustainability. The integrative nature of the Sustainable Development Goals is, in theory, supposed to address such interactions, but the complexity of the systems involved, limited knowledge, and competing interests that combine in the formulation of goals only serve to increase this challenge. Thus, one of the most important direct governance functions required will be monitoring, review, and scientific assessment. To address this requirement, governance mechanisms must not only track progress but also provide opportunities for learning both by actors about how to achieve the goals and about cause-effect relationships between goals and outcomes.

Normative Foundations

A core purpose of the Sustainable Development Goals is to provide a coherent and integrative vision for action on sustainable development. The preamble of the 2030 Agenda for Sustainable Development (UNGA 2015), which includes the Sustainable Development Goals, draws on the large body of normative text negotiated over the last 30 years to articulate this vision, including the 1992 Rio Declaration, the 2002 Johannesburg Plan of Implementation, and the "The Future We Want," the outcome document of the 2012 UN Conference on Sustainable Development. One could mine those texts, as Young and colleagues (this volume, chapter 3) have, to find the basis for a "*Grundnorm*" that encourages a holistic approach to combine a biophysical bottom line with global equity. However, the virtue of the designed ambiguity and inclusiveness of the sustainable development concept in enabling political agreement on the Sustainable Development Goals also militates against the articulation of such an underlying normative vision in the goals, even if some might hold out hope of "creatively" building on that ambiguity in the future.

In the absence of a new *Grundnorm*, there is a risk the Sustainable Development Goals mask ongoing contestation over this understanding of sustainable development, which range from complaints of marketization and the perceived commodification of nature from the left to fears of many developing country governments that any further move toward recognizing planetary boundaries in definitions of sustainable development will lead to green protectionism and undermine economic growth. Thus, one finds in the Sustainable Development Goals a call for both "sustained" and "sustainable" economic growth and employment (Goal 8), but no mention anywhere of planetary boundaries, despite being raised in negotiations for possible inclusion precisely in negotiation over the "growth" goal (*Earth Negotiations Bulletin* 2014). Similarly, justice and inclusion are included in various goals, but human rights, while mentioned in the preamble and selectively in some targets in relation to specific goals such as education or reproductive rights, do not make it into the goals.

The Sustainable Development Goals also display a tension between "balance" and "integration" of the three dimensions of sustainable development (UNGA 2012; ECOSOC 2014; Open Working Group on Sustainable Development Goals 2014, par. 9). Under such circumstances, governance institutions must somehow take advantage of what remains a contested normative vision to build processes of learning and integration that strive for greater coherence.

Scope

Squaring the mandate of the Sustainable Development Goals for integration with the wide-ranging agenda of sustainable development in practice presents an enormous governance challenge since specific goals or their components already fall within the mandate of existing intergovernmental agencies or treaty bodies, which might resist governance embedded in goal-setting at higher levels. Yet siloization is precisely what the Sustainable Development Goals are meant to counteract.

At the same time, the Sustainable Development Goals cannot do everything. Governance of the Sustainable Development Goals needs to be narrower in scope than the entire 2030 agenda, yet align with its broader vision to encompass the three dimensions of sustainable development as well as its enabling conditions. Similarly, even as the High-level Political Forum's mandate includes follow-up and review of implementation of "all the major UN conferences and summits in the economic, social and environmental fields" (UNGA 2013a, par. 7d), the call for a "focused" and "action-oriented" agenda resonates with the need for a core framing vision that Sustainable Development Goals can provide.

The Sustainable Development Goals in practice do little to mitigate this challenge, with 17 goals that range from addressing poverty eradication, employment, and equity (including gender equality), sustainable use and protection of a wide range of resources and ecosystems, to addressing food security, water and sanitation, access to energy, and climate change. Combined with the underlying normative contestation articulated in a number of the goals, focusing the agenda will be daunting.

General Purposes and Requirements for Governance of Sustainable Development Goals

Given these challenges, I argue that governance arrangements must focus on coherence and integration of Sustainable Development Goals, balance the need for high-level leadership with the mobilization of action and resources at multiple levels and diverse mixes of actors through "orchestration," and be recognized as legitimate by the community of actors engaged in pursuing the Sustainable Development Goals.

Coherence

Generically, coherence in global governance is the systematic promotion of mutually reinforcing policies that reflect legitimate social purposes. In the context of sustainable development, it therefore refers to the systematic promotion of mutually reinforcing policies across the three dimensions of sustainable development globally and at the country level (Bernstein and Hannah 2014). This definition has institutional and ideational components. Institutionally, coherence means organizations that address similar goals work synergistically instead of at cross-purposes. There ought to be means and mechanisms to learn from, coordinate, and address conflicts across institutions. Coherence also requires interinstitutional coordination to develop mechanisms for monitoring the impact of overlapping policies, assessments of progress in implementing agreed commitments or common goals, and mechanisms for addressing poor or negative performance. Ideationally, coherence means the goals or purposes of institutions reflect a common and acceptable normative framework, which should be legitimate. Coherent governance requires this normative element, since without it one could imagine perfectly institutionally coherent policies leading to undesirable ends. To the degree normative dissensus prevails, coherent governance ought to recognize tensions and trade-offs and provide mechanisms to address them.

Importantly, this understanding means coherence can be centralized or decentralized. Coherence is not primarily about sorting out the zones of overlapping competency and competition, nor does it reside in particular institutions. Coherence is a function of how rules, policies, and arrangements across dimensions of global governance are coordinated.

In the case of the Sustainable Development Goals, the focus must be on how institutional arrangements can work with the reality of a complex and fragmentary system, with the Sustainable Development Goals providing its ideational basis. Coherence also *does not* mean one-size-fits-all policies, but recognizes "diversity of contexts and challenges within and among countries" (UNTT 2012a, par. 52), an idea well entrenched in the Sustainable Development Goals. If Haas and Stevens (this volume, chapter 6) are correct that a number of goals lack normative consensus, the question is whether governance arrangements can facilitate the kind of social learning so difficult in such circumstances, and may at best need to focus on developing or bringing new knowledge to bear on learning processes and decision making.

Orchestration

The conditions of governance outlined above—an expansive mandate that requires high-level political leadership, but institutional underpinnings that provide limited direct authority over implementing actors and a lack of material resources—make orchestration the most viable strategy for governance of the Sustainable Development Goals.

Here I specifically follow Abbott et al.'s (2015a) formulation of orchestration as a governance strategy that works indirectly through other actors and organizations ("intermediaries") and uses soft modes of influence to guide and support their actions. The High-level Political Forum has the added challenge of entering an already crowded field of orchestrators engaged in sustainable development. It must, therefore, be an *orchestrator of orchestrators* that promotes coordination within a fragmented governance space. To succeed, it will require, at a minimum, high-level participation, a strong review mechanism focused on learning and improving implementation, a robust science-policy interface, and strong links with "intermediaries" within and outside the United Nations. Thus, although I focus primarily on the High-level Political Forum, it is at most just one node—albeit an important one—in a wider governance system.

Orchestration here is meant less as a musical metaphor, as in the control an arranger imposes on the parts of an orchestra, and more in the everyday meaning of directing or arranging (policies) coherently to produce desired effects. Orchestration is thus a strategy of indirect governance: the orchestrator works through intermediaries rather than attempting to govern targets directly. It contrasts with strategies in which governance actors directly engage their ultimate policy targets, whether through mandatory, hierarchical regulation or collaborative approaches such as negotiated self-regulation. Since enlistment of intermediaries will be largely voluntary in the case of the Sustainable Development Goals, it is only feasible when the policy goals of (potential) intermediaries are aligned with, or at least broadly similar to, those of the orchestrator. Orchestration is thus a strategy of soft governance; because the orchestrator lacks hard control, it must deal with intermediaries through leadership, persuasion, and incentives.

Orchestration is not a new strategy in the area of sustainable development, which has long embraced polycentricism (Ostrom 2010). Most notably, the 2002 World Summit on Sustainable Development was at the forefront of global policy in institutionalizing public-private partnerships as a primary means of implementation (Bäckstrand and Kylsäter 2014). Meanwhile, nonstate, private, and networked forms of governance have proliferated in the wake of perceived weaknesses of the multilateral system in a number of sustainable development areas (for example, Rayner, Buck, and Katila 2010; Hoffmann 2011). In other issue areas characterized by polycentrism, orchestration has been a valuable tool for increasing order and coherence without coercion, though it has also proven to be a challenging strategy (Abbott et al. 2015a, 2015b).

Legitimacy

Legitimacy will be fundamental to the ability of a lead institution to successfully orchestrate action on Sustainable Development Goals, and may have broader implications for the United Nations as a whole, given that sustainable development is evolving to be part of its core mission. The ability to orchestrate will be highly contingent on early development of legitimacy vis-à-vis the UN Economic and Social Council (ECOSOC), Bretton Woods Institutions, and the World Trade Organization (WTO), while the Sustainable Development Goals can also play a role in legitimating those institutions to the degree they are seen as pursuing policies and practices consistent with the goals.

To be legitimate and effective, governance of the Sustainable Development Goals must be sensitive to issues of ownership, buy-in, and related questions of rights and participation, as well as to conditions and enablers of implementation, including mechanisms to generate commitment and capacity. Governance of the Sustainable Development Goals must be especially sensitive to national differences and processes to avoid any perception of the goals as top-down or imposed, while simultaneously supporting and enabling progress on global goals.

The High-level Political Forum on Sustainable Development as a Lead Institution

A core lesson from 20 years of experience with the UN Commission on Sustainable Development, charged with the follow-up to the first Rio conference in 1992, is that attempts to integrate the three dimensions of sustainable development into global governance will require high-level political leadership (UN 2013a). Similarly, lessons from other attempts to mainstream crosscutting concerns, such as gender, into international institutions demonstrate the need for an institutional champion to overcome sectoral silos. In that case, the consolidation of UN entities concerned with gender equality into UN Women, with political leadership from the Commission on the Status of Women, has successfully fostered gender mainstreaming within internal operations of entities throughout the UN system. UN Women also supports strong accountability mechanisms, including UN country team performance indicators and reporting on mainstreaming in program delivery and actual development results (UN 2013b, box 9). In contrast, reforms of interagency mechanisms, such as the creation of the Chief Executives Board for Coordination and the UN Delivering as One initiative (to create coherence in program delivery), have made little progress on mainstreaming sustainable development due to a lack of political leadership (Evaluation Management Group 2012). For example, The Delivering as One pilot-phase report stated that the coordination and coherence needed to integrate the delivery of environment and development programs had not occurred (Evaluation Management Group 2012). Similarly, another coordinating body, the Environmental Management Group,[2] is still struggling to implement a UN-wide environmental and social sustainability framework (Environment Management Group 2013, 2014).

The 2012 UN Conference on Sustainable Development mandated the new High-level Political Forum to play this leadership role. It replaces the UN Commission on Sustainable Development, which had faced widespread criticism for being increasingly unable to translate discussions into action and policy impact, despite some early successes. These shortcomings stemmed from its inability to attract nonenvironmental ministers and other high-level policy makers, especially from the economic and social sectors, and consequently its weak relationships with financial, development, and trade institutions; its rigid sectoral agenda, which prevented it from addressing emerging challenges; and its limited capacity for monitoring, review, or follow-up on decisions (for example, Stakeholder Forum for a Sustainable Future 2012; UN 2013a).

In contrast, the High-level Political Forum's mandate is ambitious and expansive: setting the sustainable development agenda, including addressing emerging issues; enhancing integration, coordination, and coherence across the UN system and at all levels of governance; follow-up and review on progress in implementing all Sustainable Development Goals and commitments; providing a platform for partnerships; enhancing participation of the "major groups" and other stakeholders in decision making and implementation; and enhancing evidence-based decision making at all levels (UNGA 2012, par. 85; UNGA 2013a, par. 2, 7–8, 15, 20). It would be difficult for any institution to fulfill all of these demands on its own, yet each is important, since the integrative purpose of the Sustainable Development Goals militates against easily designating single lead agencies for each goal.

As Secretary-General Ban Ki-moon put it at the High-level Political Forum's inaugural meeting (UN Secretary-General 2013), the forum will "guide the UN system [on sustainable development] and hold it accountable." Its mandated functions—discussed below—suggest how it can do so. They are discussed in light of the governance challenges identified above, lessons learned from the experience of the UN Commission on Sustainable Development, and conditions for successful orchestration, while recognizing that many of these functions will be performed more by the High-level Political Forum's "intermediaries" than the High-level Political Forum itself. These conditions include legitimacy, focality, and political weight, and the availability of appropriate intermediaries and sufficient resources or other means to enlist their support (Abbott et al. 2015a, 2015b).

The High-level Political Forum as Orchestrator[3]

The High-level Political Forum arguably marks the first time the UN system so explicitly built orchestration into the very design of a high-level political body, with no illusion of direct hierarchical authority (Abbott and Bernstein 2015). It is empowered with considerable formal authority, but lacks direct control over resources for implementation, binding legal decision-making powers, or enforcement capability. The High-level Political Forum also enters a crowded field of orchestrators already engaged with the concerns particular Sustainable Development Goals aim to address (including UN agencies, international financial institutions, and organizations such as the Global Environment Facility, the WTO, the Group of 20 major economies, and "action networks" such as Every Women Every Child and Sustainable Energy for All). It must therefore be an orchestrator of orchestrators that promotes coordination within a fragmented system without igniting counterproductive turf wars or feeding perceptions of competition.

These conditions of "ambitious governance goals but moderate governance capacity" (Abbott et al. 2015a, 3), combined with a crowded and fluid institutional environment, are ideally suited to orchestration, since more direct or hierarchical governance modes are largely unavailable. Less positively, the High-level Political Forum also fits with another finding of orchestration research: states often initiate orchestration because it allows them to obtain a modicum of governance, including constraints on their own behavior, without having to delegate strong authority or incur high material or sovereignty costs. That is, states like orchestration because they can achieve (modest) results with weak institutions (Abbott et al. 2015b).

High-level Political Leadership: Legitimacy, Focality, and Political Weight

The High-level Political Forum is a universal body that convenes every four years under the auspices of the UN General Assembly, at the head of government level, and annually under the auspices of ECOSOC. The latter will facilitate orchestration and coordination with ECOSOC subsidiary bodies and UN agencies, as well as with regional commissions. These links, along with its universal membership, should make the High-level Political Forum highly legitimate. Its mandate also includes high-level participation and the promotion of active participation by developing countries; the UN system and other international organizations; and major groups and other stakeholders, based on practices of the UN Commission on Sustainable Development. A voluntary trust fund provides supplementary support for developing countries and stakeholders.

These features, especially the hybrid structure that links it to the General Assembly and ECOSOC, but makes it subsidiary to neither, should also make the High-level Political Forum "focal": a leader of the sustainable development policy domain. They should also give the High-level Political Forum greater "political weight" than the UN Commission on Sustainable Development. The High-level Political Forum has a mandate to set the agenda for sustainable development and integrate its three dimensions. In addition, its ministerial meetings overlap ECOSOC's high-level segment and incorporate voluntary national reviews and thematic reviews of progress on the Sustainable Development Goals.

Its ability to provide leadership, legitimacy, and access to levers of influence in national governments to mobilize commitments and resources will depend in part on the degree it can attract high-level participation. The first two High-level Political Forum meetings had notable convening success, attracting heads of governments and high-level representatives from ministries including finance, planning, children, housing, development, and foreign affairs as well as the environment, appearing to overcome the environmental bias that plagued the latter years of the UN Commission on Sustainable Development (author observations at the second session of the High-level Political Forum, 30 June–8 July 2014; *Earth Negotiations Bulletin* 2014). A clear agenda built around the Sustainable Development Goals would increase the High-level Political Forum's legitimacy and impact while facilitating action by intermediaries; thus, their relationship is synergistic.

Establishing its focality and legitimacy early on will be important if the High-level Political Forum is to influence ECOSOC, the Development Cooperation Forum, and the UN Development Group, an interagency coordinating mechanism under the Chief Executives Board for Coordination, through which development-related program delivery and implementation within the UN system is coordinated. The High-level Political Forum's influence can then also filter down to specific coordinating mechanisms such as UN Water, UN Energy, UN Oceans, or other interagency groupings that could be mandated to address specific Sustainable Development Goals. Similarly, the High-level Political Forum could act as a problem-solving forum, providing political direction on interinstitutional impasses that constrain progress on Sustainable Development Goals.

The second High-level Political Forum meeting, however, raised questions about its agenda-setting capacity and autonomy. With no independent bureau and limitations on stakeholder access, the ECOSOC leadership largely dictated the agenda of the second High-level Political Forum. Questions about access were also raised because it remains unclear whether more limited ECOSOC rules apply when it meets under its auspices, despite its mandate from the UN General Assembly to build on the modalities of participation of the more open UN Commission on Sustainable Development (Strandenaes 2014; author's observations of stakeholder interventions at the second High-level Political Forum; UNGA 2013a). Stakeholders linked these concerns to a more general worry that the High-level Political Forum's hybrid structure and "forum" status could render it weaker than its predecessor, citing as well its reduced meeting times compared to the UN Commission on Sustainable Development.

In addition, the High-level Political Forum adopted a joint ministerial declaration with ECOSOC's high-level segment as the meeting outcome (ECOSOC 2014). While this could suggest an alignment of priorities, which would be beneficial, it raises questions about the High-level Political Forum's ability to act autonomously. In addition, the Group of 77 and China tested the forum's ability to make decisions by floating a proposal on the mandated global sustainable development report.[4] Substantive divisions over the exact mandate of the report led the proponents of the proposal to withdraw it, leaving the forum's formal decision-making capacity inconclusive (*Earth Negotiations Bulletin* 2014, 15). Meanwhile, its third meeting, in 2015 under the auspices of ECOSOC, focused heavily on its future role, but with significant uncertainty, given that it fell in the midst of active negotiations on key aspects of the post-2015 agenda, including those regarding finance, governance, and follow-up (*Earth Negotiations Bulletin* 2015). Thus, it will likely be a few years before its potential to achieve focality, legitimacy, or political weight can be evaluated.

Integration and Coherence within and beyond the UN System

The High-level Political Forum's mandate gives it the potential to take over sustainable development mainstreaming, which has lacked an intergovernmental champion within the United Nations. The Sustainable Development Goals can be used as an attempt to articulate what mainstreaming might look like by identifying integrative goals and targets, which can add content to current processes of mainstreaming that have lacked both a strong normative foundation and a political champion within the system. Thus, the High-level Political Forum could, for example, provide political weight to the promotion of a UN environmental and social sustainability framework, which might appear threatening to existing agency framings or understandings of their missions. The High-level Political Forum can also invite chairs or staff of coordinating bodies to its meetings and support analyses and reviews of initiatives where integration of sustainable development (and the Sustainable Development Goals) is essential, but lacking.

The High-level Political Forum should develop a particularly strong relationship with the UN Development Group[5] to provide political guidance for the further implementation of Delivering as One to incorporate the Sustainable Development Goals. Such guidance could improve consistency, support national sustainable development planning and strategies, more effectively deliver technology support and capacity building, and administratively simplify or streamline requirements of, and support for, implementation of multilateral environment agreements and related commitments.

Creating coherence through the Sustainable Development Goals should also extend to partnerships and voluntary commitments. The High-level Political Forum and related functions of review and monitoring (see below) could help ensure the Sustainable Development Goals provide that normative coherence going forward. While integration of the three dimensions of sustainable development is already a driver of major thematic action networks, there is a need for more systematic efforts to promote coherence across the 2,110 (by latest count; see United Nations 2016a) voluntary commitments and initiatives that are the most concrete means of implementation to come out of the 2012 UN Conference on Sustainable Development.

ECOSOC, Bretton Woods Institutions, and the WTO as Intermediaries

The High-level Political Forum is explicitly mandated to develop links and invite participation of other agencies in and outside the UN system to facilitate integration and coherence. Its relationship to ECOSOC as an intermediary is complex, but crucial. ECOSOC remains "the central mechanism for coordination of the activities of the United Nations system" and for "supervision of subsidiary bodies in the economic, social, environmental and related fields" (UNGA 2013b). While fluidity between the two bodies may constrain the autonomy of the High-level Political Forum, it can also help the forum take advantage of ECOSOC's capabilities in pursuing its own mandate.

ECOSOC is also the current political platform for policy coherence across the United Nations, Bretton Woods institutions, and other economic institutions, but has had limited influence on macroeconomic issues. While there has been a rapprochement and increased participation in joint meetings and initiatives following the first financing for development conference in 2002, the high-level dialogues frequently display a UN/Bretton Woods divide. In theory, the High-level Political Forum's ability to attract economic leaders could help bridge this divide, though this too will depend on high-level participation. The High-level Political Forum's first session under the UN General Assembly successfully attracted leaders of the International Monetary Fund and World Bank; however, mostly lower-level officials from economic institutions attended its second meeting. Greater convening success may require stronger signals from ECOSOC that meetings of the High-level Political Forum are as important for coherence as its own joint meetings with the International Monetary Fund, World Bank, and WTO under the finance for development initiative.

At the same time, working relationships and orchestration among these institutions may also be facilitated through coordinating bodies and joint initiatives, although high-level political meetings can also be important for establishing cooperative mandates. For example, World Bank President Jim Yong Kim and UN Secretary-General Ban Ki-moon had agreed to accelerate the Millennium Development Goals, which had been a topic at these meetings. Initiatives on specific goals were subsequently jointly coordinated by the World Bank and UN Development Programme (UNDP) and facilitated through discussions of crosscutting issues at the Chief Executives Board for Coordination. Having these processes in place with political backing should support moving more quickly on the Sustainable Development Goals than was the case with the Millennium Development Goals.

The ECOSOC and the High-level Political Forum can also provide specific direction to the UN General Assembly on sustainable development, which in turn provides detailed guidance to member agencies. For example, its "quadrennial comprehensive policy review of UN operational activities for development" resolution provides a detailed four-year coherence agenda for the UN development system.

Links with Regional Commissions

The relationship with regional commissions should be both "up" in regard to preparatory meetings for High-level Political Forum sessions, as well as "down" in terms of follow-up and review. Regional commissions are likely to be the key conduit for supporting links of High-level Political Forum reports, declarations, or other outcomes to regional and national decision making on sustainable development policies and planning. They can also be the primary forums for bringing country views and experiences together, promote regional analyses and activities, and provide input into High-level Political Forum meetings (South Centre 2013). Regional commissions could also include their own national voluntary reviews and reporting of progress on the Sustainable Development Goals (UN TST 2013, 5–6).

Linkages to Nonstate Governance Entities

Though it will be only one node in connecting the world of business, partnerships, networks, and other implementing actors to the Sustainable Development Goals, the High-level Political Forum can facilitate learning forums, identify possible intermediaries in whatever form they occur, and then engage, support, and report on their ability to implement and scale up sustainable development. Thus, it has the potential to provide a more robust and less rigid and hierarchical understanding of partnership than those promoted following the 2002 World Summit on Sustainable Development.

Orchestration will also be important to steer commitments toward implementation of the Sustainable Development Goals, correcting a weakness of the UN Commission on Sustainable Development. The High-level Political Forum could, for example, support the UN's voluntary accountability framework and otherwise strengthen monitoring and review to measure progress and promote accountability and legitimacy. Partnerships with "precise and binding norms that are strictly monitored and enforced" perform better (Bäckstrand et al. 2012, 135). The web-based "Partnerships for SDGs" platform (United Nations 2016a) provides a promising basis to support these functions. Keeping it up to date, with ongoing input from stakeholders, and proactively linking it to other commitments and registries is essential for effective support, monitoring, collaboration, and steering (Natural Resources Defense Council and Stakeholder Forum for a Sustainable Future 2013).

The High-level Political Forum could also coordinate with UN initiatives that reach out to the private sector, such as the UN Global Compact. This approach would better foster innovation in governance and would be further recognition of the polycentric nature of sustainable development governance.

Direct Governance Functions: Science, Monitoring, and Review

A key lesson from the Millennium Development Goals is that monitoring and review processes are crucial to ensure accountability, facilitate learning, and keep pressure on implementation processes. These mechanisms are the most challenging *direct* governance functions required for effective Sustainable Development Goals.

Monitoring Progress and the Science-Policy Interface

Governments have mandated the High-level Political Forum to strengthen the science-policy interface, with a Global Sustainable Development Report being one key component (UNGA 2013a, par. 20). Ideally, the report should be forward-looking and policy-oriented, not only identifying areas of progress but also providing evidence-based analysis of policy gaps and shortcomings. Similarly, it should develop analyses that link drivers of (un)-sustainable development to outcomes, tracing how interventions and other uncertainties interact with drivers to create (un)sustainable pathways.

There are a variety of scientific and technical challenges to fulfilling these requirements. For example, difficulties of developing sustainability measures and indicators for targets for Sustainable Development Goals stem from their integrative nature, the multiple drivers of (un)sustainability, and interactions and linkages. Thus, work on sustainability scenarios should also be supported that focuses on drivers and their interactions with socioeconomic-governance factors, not just material indicators. Moreover, not all drivers are easily quantifiable. While production, consumption, and population may be quantifiable, fragility, security, and vulnerability may not be, and may have high data requirements. Despite significant progress on measurement in many areas related to sustainability—including the environment and natural resource base, climate change and biodiversity loss, and the relationships between hunger and poverty—both scientific and value assumptions about these relationships may be problematic (UNTT 2013, par. 100–101). For example, there may be political and value judgments behind the optimal relationship between, say, population, economic growth, and environmental resources. In this regard, country and

UN agency implementation reviews and lessons learned can be an important input, not only in generating data, but in feeding back into improving measures and monitoring. Such processes should also include opportunities for stakeholder involvement as well as scientific guidance, so that measures and monitoring are seen as constructive and enabling as opposed to top-down and disciplining.

These challenges also suggest that monitoring of Sustainable Development Goals should be systemic, with sensitivity to signals of systemic transition and linkages among multiple parts or processes of a system (for example, food, water, jobs, and energy when monitoring intensification of agriculture); linkages across distances; and linkages among stakeholders to understand their different interests and perspectives. Stress tests, with parallels to those employed by financial institutions, might be promoted, for example. Such monitoring will be too expensive for single organizations, so mechanisms must be put in place to collect and synthesize information from multiple sources and then organized in the spirit of learning and openness to mutual adjustment.

Administratively, the UN Statistical Commission provides guidance and advice on measurement for targets and indictors as the focal point for statistics in the UN system, but the UNDP, as it did for the Millennium Development Goals, can provide the institutional link to national-level monitoring and reporting on Sustainable Development Goals, which reflects the importance of country ownership. This governance arrangement also points to possibly different global and national objectives for monitoring: global monitoring aims to generate comparability, while the motivation for national action is ownership, policy design, and empowerment in UN processes (Mukherjee 2013). This observation resonates with studies and reports (for example, UNTT 2013; Halle, Najam, and Wolfe 2014; IISD 2014a, 2014b; *Earth Negotiations Bulletin* 2014; United Nations 2016b) on the design of review processes, which have emphasized the importance of incentives for participation, including linking reviews to building data collection and monitoring capacity at the national level, means of implementation, and lesson learning.

Science panels—especially a body or bodies that focus on synergies and integrating knowledge across sectors—may also be a way to identify knowledge gaps and interplay among sustainability challenges, as well as provide early signals of emerging sustainability threats (Haas and Stevens 2011; Haas and Stevens, this volume, chapter 6). One step in this direction is the planned creation of an "Independent Group of [15] Scientists to draft the quadrennial Global Sustainable Development Report" that will inform the High-level Political Forum when it meets under the UN General Assembly every four years (ECOSOC 2016). Improving the science-policy interface equally requires social scientific analyses of policy tools and interventions, for example in linking means of implementation such as finance, technology, and trade, with making progress on sustainable development. One important input could be the development of scenarios, or "storylines" that link drivers to outcomes, while taking uncertainties into account, thus identifying possible pathways toward sustainable development and the interaction of possible policy interventions with those pathways (UN DESA 2013).

The Global Sustainable Development Report might also include summaries of existing review and accountability reports of partnerships, voluntary commitments, and sustainable development action networks. This fits with the mandate of the report to base its findings on "information and assessments" of sustainable development, but to avoid duplication with other efforts. For example, networks such as Every Women Every Child already have a number of accountability mechanisms that report on resources and results (for a summary and links to reports of these initiatives, see Division for Sustainable Development 2013, 11–15). Assembling key findings and links to full reports in one place could not only increase transparency and accountability but also provide quick and comprehensive access to models of reporting and review, given the wide range and variety of partnership and voluntary commitments.

Review of Progress on the Sustainable Development Goals

At the time of writing, a draft UN resolution (President of the UN General Assembly 2016) had been negotiated that provides an overall approach to reviews under the auspices of the High-level Political Forum and ECOSOC. While it still leaves flexibility in how reviews will be carried out, it reflects a general consensus that review, monitoring, and accountability must be part of a larger framework, not focused only on national presentations and reviews and not only centered in the High-level Political Forum. There are also warning signs that some developing countries will resist significant strengthening of country-level review processes or deeper involvement of civil society (author's participant observation at two 2014 UN workshops sponsored by a seven-state consortium to design the High-level Political Forum review mechanism). State-led mutual review of national sustainable development progress and plans should thus be only one node in a wider system. Such reviews could use national sustainable development plans as baselines and assemble information from other reviews as part of its process. As noted above, experience with the Annual Ministerial Review also suggests the need to develop incentives and support for participation, focus on learning opportunities, and lead to calls for correction, coherent action, and means of implementation when gaps are discovered (IISD 2014a, IISD 2014 b; United Nations 2016b). The quadrennial UN General Assembly meetings of the High-level Political Forum could even be an opportunity to consider revisions or modifications of the targets and indicators for Sustainable Development Goals as new knowledge becomes available.

The High-level Political Forum is also in a unique position to promote review, accountability, and learning for action networks, partnerships, and voluntary commitments. For example, it could encourage independent third-party reviews, facilitated by emerging tools and platforms both inside the UN system (such as the partnership platform) and outside. However, active encouragement and some technical and material support will be needed to establish these practices and robust platforms, especially to facilitate participation by stakeholders from developing countries and more marginalized major groups.

Mobilization of Means of Implementation

Progress on the Sustainable Development Goals will require continuing entrepreneurship and reliable, predictable resource mobilization. Implementation of goals can be enhanced when regional organizations and commissions, countries, provinces, and municipalities generate and receive technical and scientific inputs and support to inform sustainable development stakeholder engagement and activities on multiple governance levels.

Partnerships, action networks, and transnational actors, including non-state sustainability standard setters along marketplace supply chains, will also be crucial players. The voluntary commitments made at and since the 2012 UN Conference on Sustainable Development account for the vast majority of financial and other resources thus far for the means of implementation of the Sustainable Development Goals, though obviously countries' own domestic resource mobilization and policy commitments will be crucial as well. The High-level Political Forum can also promote new initiatives, much the way the Muskoka Summit of the Group of Eight major economies launched the Muskoka Initiative for Maternal, Newborn and Child Health, expected to mobilize more than US$10 billion. However, the US$636 billion figure cited after 2012 for these commitments does not differentiate existing from new commitments, nor is there yet an accountability mechanism to ensure that the commitments fit with the Sustainable Development Goals.

Earlier experiences with partnerships relevant to the Millennium Development Goals and sustainable development more broadly show that lack of institutionalized review mechanisms and of clear, quantifiable benchmarks to measure performance contributed to their uneven effectiveness (Bäckstrand et al. 2012, 133–141). Moreover, partnerships "have a poor record of promoting systemic change," while "partnerships with a focus on specific, short-term quantifiable results can also detract funding from long-term investment essential to promote long-term development" and, if they create "separate parallel structures," can weaken country ownership (UNTT 2013, 8).

Sustainable Development Goals also cannot be successfully implemented without government commitments, long-term investments, and new sources of funding. While official development assistance from OECD countries and from new non-Western donors is important for addressing poverty and other global issues, there will need to be greater reliance on development finance from private-sector investments, support from non-governmental organizations and foundations, and domestic resource mobilization (see Voituriez et al., this volume, chapter 11). However, investment must be channeled to sustainable development generally, and in particular to low-carbon technologies, green growth, and infrastructure development. Investments must also be made in inclusive development, even if risk-reward ratios and long time frames traditionally place them "outside the investment parameters" of many long-term investors (ECOSOC 2013; see also Intergovernmental Committee of Experts on Sustainable Development Financing 2014).

Conclusion

This chapter is at once more and less optimistic than Underdal and Kim's, which similarly focuses on orchestration. On the one hand, it is less optimistic that orchestration can arise among existing institutions absent significant and active governance support, even if the SDGs evolve to meet all the conditions for effective goal setting (which, both chapters agree, is unlikely). Slightly more optimistically, here I have tried to identify institutional means to support goal attainment and conditions under which they can catalyze the necessary actions, commitments, and resources to move toward goal attainment.

To say this chapter is only "slightly" more optimistic, however, is to share their caution in overstating the difference institutional arrangements can make for effective goal attainment with the means currently available. What I have tried to do is identify important institutional purposes and mechanisms, and the conditions under which they have the best likelihood of mobilizing action on the Sustainable Development Goals. A number of risks and cautions are warranted in this regard.

First, the United Nations itself is not without its own legitimacy challenges, especially related to questions of efficacy and "focality" vis-à-vis minilateral arrangements such as the Group of 20 major economies or new governance arrangements outside of official multilateral processes. Even if the Sustainable Development Goals generate publicity, motivate action, and generate useful benchmarks, a lead institution closely tied to ECOSOC could face legitimacy challenges or, even if ostensibly viewed as an appropriate focal point for governance, be ignored by governments, especially in the North, skeptical of its ability to influence actors or organizations with significant resources and direct levers to make the most progress on the Sustainable Development Goals: the Bretton Woods Institutions, WTO, the Group of 20, and the private sector.

Second, while consultations on the environment and the post-2015 agenda generated an "overwhelming call for environmental sustainability to be at the heart" of that agenda, the evidence so far is that the same North-South political dynamics that produced tensions between "balance" and "integration" in the 17 Sustainable Development Goals are being mirrored in the High-level Political Forum. The experience of Millennium Development Goal 7 is cautionary, where performance on environmental targets "have been particularly poor" (UNEP and UNDP 2013). Thus, if the High-level Political Forum is to make a difference, the engagement of UN Environment Programme (UNEP) and its technical capacity, as well as other forums that have had more success in integrative policy thinking, is imperative for mainstreaming sustainable development within the UN system, building links to multilateral environmental agreements, and more broadly for transformative change toward the sustainability that the Sustainable Development Goals are supposed to catalyze.

Finally, there is a risk that the broader context to the "global partnership for sustainable development" of Goal 17 is being lost in the emphasis on voluntary commitments and partnerships as a primary means of implementation. The publicity for the goals even shortens Goal 17 to "Partnerships for the Goals" (e.g., on the main goals website, http://www.un.org/sustainabledevelopment/sustainable-development-goals/). This focus risks giving insufficient attention to development finance and trade rules, including concerns about market access, subsidies, preferential regional trade agreements, and funding for "aid for trade," all of which were highlighted by the Intergovernmental Committee of Experts on Sustainable Development Financing (2014) in its final report on the post-2015 development agenda. The partnership for development identified in Millennium Development Goal 8 focused on these issues, a framing less centered on voluntary commitments for implementation than the current formulation of Goal 17. Thus, although Sustainable Development Goal 17 contains targets that include structural concerns around finance, aid, trade, and technology transfer, it adds ambiguity to the meaning of "partnership" by including explicit references to multi-stakeholder partnerships and domestic resource mobilization as central elements of the "means of implementation" and the "global partnership for sustainable development." It is important in this regard that monitoring and indicators of the Sustainable Development Goals continue to shed light on how trade rules and policies, for example, affect trade flows, market access, and technology transfer. There is a role for the High-level Political Forum in following up and reviewing Sustainable Development Goals to highlight specific issues like access to essential medicines, market access for environmental goods and services, and work on social and environmental standards, as well as generally improving coherence of sustainable development in the marketplace. It at least has the potential to remedy Millennium Development Goal 8's former lack of an overarching political forum to have a strategic discussion about implementation. The High-level Political Forum could then nudge other forums and organizations responsible for negotiations and implementation of the partnership for development or standard setting in the area of sustainable development, both governmental and nongovernmental, toward best practices and more coherent policies.

Notes

1. For a preliminary attempt, see UN TST (2013, 2), which summarizes Sumner and Tiwari (2010) and UNTT (2012b).

2. Its membership is similar to the Chief Executives Board for Coordination, but it focuses specifically on coordinating efforts on environmental issues.

3. This subsection on orchestration and those that follow on the functions and role of the High-level Political Forum draw liberally from detailed discussions in Bernstein (2013) and Abbott and Bernstein (2015).

4. Proponents cited UN General Assembly resolution 67/290, par. 9, which provides that the High-level Political Form operates under the same rules of procedure as functional commissions when meeting under ECOSOC auspices.

5. Arguably, renaming it the UN Sustainable Development Group would better reflect the move to a more universal and integrated 2030 agenda, but no such proposal is currently being considered.

References

Abbott, Kenneth W., and Steven Bernstein. 2015. The High-level Political Forum on Sustainable Development: Orchestration by Default and Design. *Global Policy* 6 (3): 222–233.

Abbott, Kenneth W., Philipp Genschel, Duncan Snidal, and Bernhard Zangl. 2015a. Orchestration: Global Governance through Intermediaries. In *International Organizations as Orchestrators*, ed. Kenneth W. Abbott, Philipp Genschel, Duncan Snidal and Bernhard Zangl, 3–36. Cambridge, UK: Cambridge University Press.

Abbott, Kenneth W., Philipp Genschel, Duncan Snidal, and Bernhard Zangl. 2015b. Orchestrating Global Governance: From Empirical Findings to Theoretical Implications. In *International Organizations as Orchestrators*, ed. Kenneth W. Abbott, Philipp Genschel, Duncan Snidal, and Bernhard Zangl, 349–379. Cambridge, UK: Cambridge University Press.

Bäckstrand, Karin, Sabine Campe, Sander Chan, Ayşem Mert, and Marco Schäferhoff. 2012. Transnational Public-Private Partnerships. In *Global Environmental Governance Reconsidered*, ed. Frank Biermann and Philipp Pattberg, 123–147. Cambridge, MA: MIT Press.

Bäckstrand, Karin, and Mikael Kylsäter. 2014. Old Wine in New Bottles? The Legitimation and Delegitimation of UN Public-Private Partnerships for Sustainable Development from the Johannesburg Summit to the Rio +20 Summit. *Globalizations* 11 (3): 331–347.

Bernstein, Steven. 2013. The Role and Place of a High-level Political Forum in Strengthening the Global Institutional Framework for Sustainable Development. Commissioned by UN Department of Economic and Social Affairs. Available at: http://sustainabledevelopment.un.org/content/documents/2331Bernstein%20study%20on%20HLPF.pdf.

Bernstein, Steven, and Erin Hannah. 2014. Coherence and Global Sustainable Development Governance. Paper presented at the 2014 International Studies Association Convention, Toronto, March 26–29, 2014.

Division for Sustainable Development. 2013. Sustainable Development in Action. Special Report: Voluntary Commitments and Partnerships for Sustainable

Development. SD in Action Newsletter, 1. Prepared by the Outreach and Communications Branch, UN Department of Economic and Social Affairs.

Earth Negotiations Bulletin. 2014. Summary of the Second Meeting of the High-level Political Forum on Sustainable Development, June 30–July 9, 2014. *Earth Negotiations Bulletin* 33 (9).

Earth Negotiations Bulletin. 2015. Summary of the 2015 Meeting of the High-level Political Forum on Sustainable Development, June 26–July 8, 2015. *Earth Negotiations Bulletin* 33 (18).

ECOSOC, UN Economic and Social Council. 2013. Summary by the President of the Economic and Social Council of the special high-level meeting of the Council with the Bretton Woods institutions, the World Trade Organization and the United Nations Conference on Trade and Development (New York, 22 April 2013) (A/68/78–E/2013/66).

ECOSOC, UN Economic and Social Council. 2014. High-level Segment and High-level Political Forum on Sustainable Development: Adoption of the Ministerial Declaration of the High-level Political Forum. UN Doc. E/2014/L.22–E/HLPF/2014/L.3.

ECOSOC, UN Economic and Social Council. 2016. Zero-Draft: Ministerial Declaration of the 2016 High-level Political Forum on Sustainable Development convened under the auspices of the Economic and Social Council on the theme "Ensuring than no one is left behind."

Environment Management Group. 2013. System-wide Issues in the Follow up of the Framework for Advancing Environmental and Social Sustainability in the UN system. Options paper. Available at: http://www.unemg.org/images/emgdocs/safeguards/131129%20-%20ess%20draft%20option%20papers.pdf.

Environment Management Group. 2014. Advancing the Environmental and Social Sustainability Framework in the United Nations System: Interim Guide. Geneva: United Nations.

Evaluation Management Group. 2012. Independent Evaluation of Delivering as One: Summary Report. New York: United Nations.

Haas, Peter M., and Casey Stevens. 2011. Organized Science, Usable Knowledge and Multilateral Environmental Governance. In *Governing the Air*, ed. Rolf Lidskog and Göran Sundqvist, 125–161. Cambridge, MA: MIT Press.

Halle, Mark, Adil Najam, and Robert Wolfe. 2014. *Building an Effective Review Mechanism: Lessons for the HLPF*. International Institute for Sustainable Development.

Hoffmann, Matthew J. 2011. *Climate Governance at the Crossroads: Experimenting with a Global Response after Kyoto*. New York: Oxford University Press.

IISD, International Institute for Sustainable Development. 2014a. Options for the High-level Political Forum Review Mechanism: Background for the Second Workshop. May 15, 2014, New York.

IISD, International Institute for Sustainable Development. 2014b. A Briefing Note of the Second Workshop on "Making the High-level Political Forum on Sustainable Development Work: How to Build an Effective 'Review Mechanism.'" *HLPF Bulletin* 221 (2). Available at: http://www.iisd.ca/hlpf/hlpfsdw2.

Intergovernmental Committee of Experts on Sustainable Development Financing. 2014. Report of the Intergovernmental Committee of Experts on Sustainable Development Financing. UN Doc. A/69/315. Available at: http://www.un.org/ga/search/view_doc.asp?symbol=A/69/315&Lang=E.

Mukherjee, Shantanu. 2013. Learning from National MDG Reports. Presentation to the Open Working Group, New York.

Natural Resources Defense Council and Stakeholder Forum for a Sustainable Future. 2013. Fulfilling the Rio+20 Promises: Reviewing Progress since the UN Conference on Sustainable Development. Available at: http://www.nrdc.org/international/rio_20/.

Open Working Group on Sustainable Development Goals. 2014. Outcome document. July 19. Available at: http://sustainabledevelopment.un.org/focussdgs.html.

Ostrom, Elinor. 2010. Polycentric systems for coping with collective action and global environmental change. *Global Environmental Change* 20 (4): 550–557.

President of the UN General Assembly. 2016. Letter from the co-facilitators of the General Assembly informal consultations on the 2030 agenda follow-up (Belize and Denmark) and a zero draft resolution (6 May 2016). Available at: https://sustainabledevelopment.un.org/content/documents/10124Follow-up%20and%20review%20-%206%20May%202016.pdf.

Rayner, Jeremy, Alexander Buck, and Pia Katila, ed. 2010. Embracing Complexity: Meeting the Challenges of International Forest Governance. A Global Assessment Report Prepared by the Global Forest Expert Panel on the International Forest Regime. Vienna: International Union of Forest Research Organizations.

South Centre. 2013. Concept paper by South Centre on High-level Political Forum on Sustainable Development. Geneva.

Stakeholder Forum for a Sustainable Future. 2012. Review of Implementation of Agenda 21 and the Rio Principles: Synthesis. New York: United Nations Department of Economic and Social Affairs.

Strandenaes, Jan-Gustav. 2014. Participatory Democracy—HLPF Laying the Basis for Sustainable Development Governance in the 21st Century. Report for UN Department of Economic and Social Affairs. Available at: http://sustainabledevelopment.un.org/index.php?menu=1564.

Sumner, Andy, and Meeta Tiwari. 2010. Global Poverty Reduction to 2015 and Beyond: What Has Been the Impact of the MDGs and What Are the Options for a Post-2015 Global Framework? Institute of Development Studies at the University of Sussex. Working paper 348.

UN, United Nations. 2013a. Lessons Learned from the Commission on Sustainable Development: Report of the Secretary-General. UN General Assembly. UN Doc. A/67/757.

UN, United Nations. 2013b. Report of the Secretary-General on Mainstreaming of the Three Dimensions of Sustainable Development throughout the United Nations System. UN General Assembly and ECOSOC. 8 May. Advance Unedited Copy. UN Doc. A/68.

UN, United Nations. 2016a. Partnerships for SDGs. Available at: https://sustainabledevelopment.un.org/partnerships.

UN, United Nations. 2016b. Critical Milestones towards Coherent, Efficient and Inclusive Follow-up and Review at the Global Level. Report of the Secretary-General. UN Doc. A/70/684.

UN DESA, United Nations Department of Economic and Social Affairs. 2013. Executive Summary of the Prototype Global Sustainable Development Report. Available at: http://sustainabledevelopment.un.org/content/documents/975GSDR%20Executive%20Summary.pdf.

UNEP and UNDP, United Nations Environment Programme and United Nations Development Programme. 2013. The Global Thematic Consultation on Environmental Sustainability in the Post-2015 Development Agenda. Draft Final Report, July 2013.

UNGA, United Nations General Assembly. 2012. The Future We Want. Rio+20 United Nations Conference on Sustainable Development, Rio de Janeiro, Brazil. UN Doc. A/CONF.216/L.1 adopted by the UN General Assembly in Res. 66/288.

UNGA, United Nations General Assembly. 2013a. Format and Organizational Aspects of the High-level Political Forum on Sustainable Development. UN Doc. A/RES/67/290.

UNGA, United Nations General Assembly. 2013b. Review of the Implementation of General Assembly Resolution 61/16 on the Strengthening of the Economic and Social Council. UN Doc. A/RES/68/1.

UNGA, United Nations General Assembly. 2015. Transforming Our World: The 2030 Agenda for Sustainable Development. Draft resolution referred to the United Nations summit for the adoption of the post-2015 development agenda by the General Assembly at its sixty-ninth session. UN Doc. A/70/L.1.

UN Secretary-General. 2013. Secretary-General's remarks at inaugural meeting of the High-Level Political Forum on Sustainable Development, New York, September 24, 2013. Available at: http://www.un.org/sg/statements/index.asp?nid=7122.

UN TST, United Nations Technical Support Team. 2013. Global Governance. Issues brief. Available at: http://sustainabledevelopment.un.org/content/documents/2429TST%20Issues%20Brief_Global%20Governance_FINAL.pdf.

UNTT, United Nations System Task Team on the Post 2015 UN Development Agenda. 2012a. Realizing the Future We Want for All: Report to the Secretary-General. New York. Available at: http://www.un.org/millenniumgoals/pdf/Post_2015_UNTTreport.pdf.

UNTT, United Nations System Task Team on the Post 2015 UN Development Agenda. 2012b. Review of the Contributions of the MDG Agenda to Foster Development: Lessons for the Post-2015 UN Development Agenda. Discussion Note.

UNTT, United Nations System Task Team on the Post 2015 UN Development Agenda. 2013. Statistics and Indicators for the post-2015 Development Agenda. New York. July. Available at: http://www.un.org/en/development/desa/policy/untaskteam_undf/UNTT_MonitoringReport_WEB.pdf.

10 The Sustainable Development Goals and Multilateral Agreements

Arild Underdal and Rakhyun E. Kim

This chapter builds on Young's insightful analysis (this volume, chapter 2) of potential advantages and limitations of goal setting as a general strategy for earth system governance, but focuses on one particular function, namely that of reforming or rearranging *existing* elements of earth system governance so as to enhance overall performance in promoting sustainable development.

Broad agreement on a new set of goals for global governance can, at least for a while, open new windows of opportunity for innovative change, including the establishment of *new* rules and institutions. However, at least in a five- to 10-year perspective, the impact of the Sustainable Development Goals will depend primarily on their success in being actively pursued by existing institutions. (In this chapter, we use the label of "institution" broadly, referring to organizations as well as regimes and international agreements.) Most of these institutions will be deeply immersed in their own agendas and will be valued by their members primarily for pursuing the missions for which they were established. A similar observation could apply to several other goal-setting attempts as well, but the Sustainable Development Goals stand out as particularly demanding in their ambitions to provide guidance *across* established policy domains. The notion of sustainability itself has implications for a very wide range of human activities, and so does the notion of development. To be effective, the Sustainable Development Goals must therefore penetrate or in some other way bring into line *existing* regimes and organizations, in particular powerful institutions established (primarily) for other purposes.

The 2030 Agenda for Sustainable Development includes 17 Sustainable Development Goals, each with a set of more specific targets. With such a comprehensive agenda, the implementation of the Sustainable Development Goals will depend critically on how thousands of agents already engaged in governing human affairs—ranging from local councils and national governments to international organizations such as the World Bank, the International Monetary Fund, and the World Trade Organization—respond within their respective domains.

Enthusiasts and optimists will likely see the 2030 Agenda for Sustainable Development as constituting an overarching and compelling normative framework for governance at all levels and within all relevant policy domains. Pessimists—probably themselves preferring the label "realists"—will likely point out that (i) several of the goals (for example, "ensure healthy lives and promote well-being for all at all ages"), and even more so the declaration at large, provide scant guidance for prioritizing scarce resources; and (ii) no hierarchically integrated world governance system exists that can enable and enforce universal and effective implementation. In the absence of clear priorities and an integrated system of world governance, the impact of the Sustainable Development Goals on existing regimes and organizations will depend heavily on the normative clout of the goals themselves and whatever soft modes of influence can be exercised by their supporters.

The word "orchestration" originated (in the nineteenth century) from the word "orchestra" as a label for a particular kind of activity, namely "the arrangement of a musical composition for performance by an orchestra," according to *Merriam-Webster*. This is a definition most of us would immediately recognize and probably also subscribe to. More recently, however, the label has come to be used also in a broader sense, as a general reference to efforts at arranging different elements of a system in harmony with each other to enhance their collective performance. In applying the label orchestration to the Sustainable Development Goals, we adopt the latter interpretation. More precisely, we conceive of orchestration as a soft mode of (global) governance, one involving efforts to induce, facilitate, and coordinate voluntary contributions to a common cause (Abbott and Snidal 2009; Abbott and Snidal 2010; Abbott et al. 2015; see also Bernstein, this volume, chapter 9). Everything else constant, the need for coordination tends to increase with the complexity and fragmentation of the system to be governed (Zelli and van Asselt 2013). Applying this proposition to the task in focus here, the Sustainable Development Goals clearly stand out as a truly demanding encounter between an ambitious and rather loosely integrated declaration of goals and a complex and fragmented system of global governance.

Like all other activities, orchestration requires one or more agents ("orchestrators"). In their general analysis, Abbott et al. (2015) point to intermediaries (often nongovernmental organizations) as the main operators in the field; in fact, working through intermediaries to govern a third set of actors (the targets) in pursuit of the orchestrator's goals is one of their defining characteristics of orchestration. Here, we simply adopt Bernstein's (this volume, chapter 9) assumption that the High-level Political Forum on Sustainable Development will be the principal UN orchestrator, and thus serve as "an orchestrator of orchestrators." The High-level Political Forum is a *forum* where ministers and (every fourth year) heads of state or government meet to provide guidance and feedback through negotiated and broadly consensual statements. What it can provide is a mode of orchestration relying mainly on rational deliberation and soft power (see, for example, Risse 2000; Nye 2004). This kind of orchestration can no doubt provide useful contributions, but even if the High-level Political Forum successfully performs its role as the global forum for guidance and feedback, it will only marginally enhance the overall capacity of the UN system to *implement* the wide range of measures required to achieve the Sustainable Development Goals. The tasks of goal *setting* and goal *attainment* differ in important respects, and the Sustainable Development Goals agenda is clearly a case where the latter task will be far more demanding than the former.

This chapter proceeds as follows. In section 2, we ask: What are the political and entrepreneurial challenges peculiar to orchestrating contributions from *existing* international institutions? In section 3, we ask: *Under which conditions* can goal setting be an effective tool for this particular task? This section builds in part on Young's analysis of "the nature of the problem" and "features of the setting" (this volume, chapter 2). In section 4, we ask: *To what extent* do the Sustainable Development Goals, as far as can be judged at this stage, meet these conditions? Section 5 summarizes our conclusions.

Orchestrating Existing International Agreements and Organizations

Since arranging music for an orchestra and arranging governance for the world are very different activities, making any direct analogy between the two would be farfetched. A brief compare-and-contrast review may nonetheless help us better understand what makes the task of orchestrating contributions from existing international institutions a particularly demanding challenge.

A philharmonic orchestra is a large ensemble of musicians (at full size, often around a hundred), each highly specialized in one particular type of instrument. Some of these musicians may well hold divergent views on what the orchestra's overall repertoire should be, or the artistic interpretation of a particular piece of music. These and other issues may generate intense discussions that in some cases enhance performance, and in other cases go sour. Even when perfectly professional and constructive, such internal discussions will have to be closed, or at least suspended, by the time the orchestra goes on stage. Once facing their audience, everyone will be involved in a *joint* project guided by a common goal, a common script, and a common interpretation of that script, communicated through the gestures of the conductor. Their interaction will be very intense, but only for limited periods of time. Because musicians perform in a setting where they can easily communicate with each other and share a sense of unity, some orchestras could, and smaller ones often do, work well in the absence of a conductor (Khodyakov 2007). In fact, full-size philharmonic orchestras that are led by a conductor also make extensive use of self-synchronization mechanisms such as a clearly specified division of labor and, in concert, continuous mutual adjustments to achieve harmony. In the end, the orchestra will be evaluated primarily for its *collective* performance, and precise coordination within as well as across groups of musicians and categories of instruments will be a necessary condition for a high rating.

Global governance constitutes a very different setting. The Sustainable Development Goals involve other kinds of *agents* (institutions more than individuals), other kinds of *activities* (governance rather than music), and multiple *scales* (in a temporal as well as in a spatial sense). Moreover, while an orchestra is constituted as one single organization, the world community is governed by a highly fragmented system. These and other contrasts imply that it would be preposterous to expect the Sustainable Development Goals to match the very high level of precise coordination required in the sphere of music; measured in terms of precise instructions, Bach's *Christmas Oratorio* and the declaration of Sustainable Development Goals are worlds apart. The bright side is that global governance can make truly important contributions to sustainable development without achieving perfect orchestration.

The current "population" of international regimes and organizations has developed over more than a century. The establishment and design of each institution can be seen as a collective response to a specific governance challenge perceived as important by its members at a particular point in time. Thus, each institution is charged with performing a limited range of specific functions for its members, and members evaluate an institution primarily in terms of how well it performs these functions. In many cases, institutional mandates, designs, and membership change over time in response to changes in task environments or as a consequence of internal disputes or bad failures. Inherent in most institutions, however, are mechanisms working to preserve their "essence" as defined at their creation and through directional decisions made at critical junctures throughout their history (Allison 1971; Pierson 2000). These mechanisms normally generate a tendency to approach tomorrow's challenges through yesterday's lessons and by means of today's repertoires. In some instances, however, path dependency turns into institutional sclerosis, a more malignant lock-in that severely impairs an institution's ability to rationally adapt to changing circumstances (Olson 1982).

As a consequence of its (co-)evolutionary development, the global governance system existing today is highly fragmented in terms of issue areas, specific mandates, memberships, and geographical range. This multidimensional fragmentation has led to increasing calls for reforms that build mutually supportive relationships among international institutions in related fields (Sanwal 2004; Pavoni 2010). For example, in the area of international environmental governance, a number of proposals have been put forward to better coordinate existing multilateral agreements and improve their collective performance. Most of these proposals are organizational in nature, such as clustering agreements with the possibility of co-locating their secretariats (Oberthür 2002; von Moltke 2005) or upgrading the UN Environment Programme into a World Environment Organization (Biermann and Bauer 2005). Such reforms can facilitate learning, enhance managerial capacity and efficiency, and in other ways improve the (technical) performance of treaty bodies such as secretariats. They do, however, fall short of aligning specific treaty rules and objectives with a common purpose.

The institutional fragmentation is particularly pervasive in certain policy domains, such as the environment, where there is a plethora of international agreements, but there are no rules or principles with a *jus cogens* status universally applicable irrespective of state consent (Boyle 2007). In the absence of an overarching collective goal that can help pull the system together, international environmental law and governance have not been sufficiently more than the sum of agreements (Kim and Bosselmann 2013). This observation applies equally to international law and governance of sustainable development (Pauwelyn 2003). The good news is that sustainable development has been increasingly accepted as a guiding concept and pursued as an ultimate objective of the international community at large. Some commentators express a degree of optimism that this collective goal of sustainable development will be instrumental in addressing the fragmentation problem by integrating different areas of international law and governance (for example, Weeramantry 2004; Voigt 2009).

Under Which Conditions Can Goal Setting Be an Effective Tool for Orchestration?

Goal setting can be a useful tool for enhancing earth system governance, but only under certain conditions. Our interest in better understanding these contextual conditions stems in part from the fact that the UN Millennium Declaration adopted by the UN General Assembly in 2000 included a commitment to "ensure greater policy coherence and better cooperation ... with a view to achieving a fully coordinated approach to the problems of peace and development" (UNGA 2000, par. 30). Although some instances of progress may be identified, the overall fragmentation of international law and governance seems not to have been significantly reduced since that declaration was adopted. Can we reasonably expect the UN system to achieve "a fully coordinated approach," or at least do better, when it comes to sustainable development?

At least three conditions seem necessary for goal setting to serve as an effective tool for orchestrating sustainable development governance (see also Bernstein, this volume, chapter 9). First, the decision-making bodies involved must be able to *reach agreement* on a small and manageable set of Sustainable Development Goals. In this case, the contextual conditions under which goal setting takes place are clearly not favorable. The world community is characterized by, among other things, stark asymmetries between rich and poor, intense competition over wealth and power, and weak institutional capacity (with consensus as the default decision rule). Under these conditions, actor preferences and beliefs are likely to diverge significantly, leaving global conference diplomacy with very demanding challenges of integration and aggregation. As a consequence, package deals and formulations deliberately designed to allow more than one interpretation ("creative ambiguity") often are important components of global agreements.

Second, the goals must *provide clear guidance* for "agents" as well as for their "principals." Agents have different substantive priorities or pursue goals of their interest (Abbott et al. 2015). Orchestration can be understood as an effort to align these goals with those of the principals. Ideally, therefore, governance through goal setting should provide a clearly specified, internally coherent, and hierarchically ordered goal structure. For reasons given above, this is far more than we can legitimately expect. The longer and more diverse the list of goals, the weaker will be its capacity to serve as a tool for orchestrating existing institutions. One less ambitious achievement that could enhance orchestration capacity would be agreement on *one single, overarching goal* as superior to all other complementary and auxiliary goals. Such a superior goal would act as a *"Grundnorm"* that gives all relevant international regimes and organizations a shared primary purpose to which they must contribute (see Young et al., this volume, chapter 3). Over time, its influence could be enhanced through, among other things, active demand and support from nongovernmental organizations or through international legal proceedings, reinforcing a planetary boundaries interpretation of sustainable development (Rockström et al. 2009) as "a principle with normative value" (International Court of Justice 1997, 85).

The experience from the Millennium Development Goals provides an instructive illustration of the challenges to be faced in the absence of such an overarching goal. Despite all being well intended, different Millennium Development Goals have sometimes led to contradictory outcomes; that is, outcomes that improve the performance of one goal by degrading another. The UN Environment Programme notes in its recent report that the current way of reducing poverty and hunger (Millennium Development Goal 1) is linked to "greater output from modern agriculture accompanied by its requirements for water, synthetic chemical fertilizers, herbicides and intensive use of machinery," thereby having a negative impact on environmental sustainability (Millennium Development Goal 7; UNEP 2013, 11). Conflicts may also be observed among environmental targets and indicators within Goal 7 itself. For example, expanding biofuel crop plantations, while potentially contributing to reductions in "CO_2 emissions, total, per capita and per $1 GDP" (Indicator 7.2), will likely decrease the "proportion of land area covered by forest" (Indicator 7.1; Danielsen et al. 2008; Fargione et al. 2008). Likewise, replacing hydrochlorofluorocarbons with hydrofluorocarbons that have zero ozone-depletion potential, while contributing to reducing the level of "consumption of ozone-depleting substances" (Indicator 7.3), exacerbate climate change because hydrofluorocarbons have a high global warming factor (Velders et al. 2007).

Considered as a stand-alone commandment, the UN Millennium Declaration goal of "achieving a fully coordinated approach" may give sufficient direction for the principal (notably the UN Secretary-General). For lower-level agencies and other actors charged with more narrowly defined mandates, however, this goal would have to be translated into behavioral guidelines referring to their particular domains. In a setting characterized by stark asymmetries, strong competition, and weak institutional capacity, this is no easy task. Moreover, institutional arrangements are *tools*, valued primarily for their contributions to achieving other goals. The UN Millennium Declaration includes multiple substantive goals—such as eradicating extreme poverty and hunger and ensuring environmental sustainability—that stand on their own merit and presumably rank higher in the order of political priorities. In competition over scarce human and financial resources, the latter type of goal is likely to prevail.

These examples of competition between goals referring to only one or a few components of the complex earth system highlight the importance of a systems approach to goal setting. Identifying crosscutting (or integrated) goals that do not end up with a false opposition between human development and planetary must-haves is essential (Griggs et al. 2013; UNEP 2013). In an era of growing complexity, it is increasingly becoming necessary to embrace trade-offs between different goals and targets that are inherent in the system itself. The threats addressed and the solutions outlined by individual governance efforts have to be evaluated against overall performance. Ideally, an overarching goal would act as a point of reference for orchestration of existing international institutions by building systemic relationships and managing trade-offs between Sustainable Development Goals with a view to ensure *overall* improvements in terms of sustainable development (Costanza et al. 2015).

Third, the agents involved must be *willing and able* to work effectively together to achieve the goal(s) set for them. Willingness may build on internalization of underlying values and norms or on external "carrots" (incentives) and "sticks" (penalties). As pointed out above, internalization of and synchronization to a team goal can, at least in some contexts, be an effective mechanism of voluntary, bottom-up orchestration. However, students of organizational behavior point out that (large) organizations respond to complexity by differentiating tasks internally and drawing boundaries with other organizations (see for example, Scott 1981). A highly specialized unit, including members and staff, tends to focus primarily on its own particular domain and to develop ideas about the "essence" of its mission and its "proper" repertoire of programmatic activities (Allison 1971). These "local" ideas and routines are more likely to be internalized than are high-level policy declarations emanating from some remote global conference. Moreover, in some instances, calls for "full" or extensive coordination with other agencies or units will most likely be perceived as threatening. This is particularly true in the international context, where efforts by actors in one regime to influence rule development or implementation in another with a different membership are often perceived as a threat to national sovereignty (Wolfrum and Matz 2003). For example, an important barrier to cooperation between the UN Framework Convention on Climate Change and the Convention on Biological Diversity has been that the United States is party to the former but not to the latter. The United States has emphasized that each convention has "a distinct legal character, mandate and membership" and insisted that biodiversity issues be dealt with outside the climate regime (UN Framework Convention on Climate Change 2006, 16). It has been repeatedly stressed, therefore, that any attempt at enhancing institutional coordination must at the same time respect the legal autonomy of the treaties (for example, UNGA 2005, par. 169). In such cases, orchestration will likely have to involve measures that would encourage or bind institutions to pursue a high-level, "suprainstitutional" goal.

An organization's *ability* to contribute is primarily a function of its human, institutional, and material resources. Contributing to achieving a new target will often require new expertise (perhaps also rearrangement of existing sections or teams), procedural and perhaps other institutional reforms, and—at least as perceived by the units most directly involved—additional funding. Even where it seems obvious that reducing duplication of work and tapping synergy potentials *can* cut costs significantly and thus make funds available for more important work, "parochial" concerns often block or at least dilute cost-saving measures. As a consequence, the net gains actually achieved often fail to meet expectations.

Do the Sustainable Development Goals Meet These Conditions?

The organization of the 2030 Agenda for Sustainable Development containing the Sustainable Development Goals resembles that of the UN Millennium Declaration. The most remarkable difference is the increase in the number of goals from eight to 17, which came about despite the agreement at the 2012 UN Conference on Sustainable Development that the Sustainable Development Goals "should be ... limited in number" (UNGA 2012, par. 247). This increase may, in part, be a function of the greater complexity of the task at hand; achieving *sustainable* development is in important respects a more demanding challenge than "merely" achieving *development*. The larger goal set probably also reflects the limited capacity of global conference diplomacy, including the Open Working Group on Sustainable Development Goals, to integrate diverging views and to aggregate conflicting preferences. In fact, a number of goals have been suggested to lack consensus (Haas and Stevens, this volume, chapter 6). These two factors—greater complexity and failed diplomacy—likely interact synergistically; as students of international environmental regimes have pointed out, the success rate tends to decline sharply when politically malignant problems are to be handled by institutions with low integration and aggregation capacity (see for example, Underdal 2002). In any case, the substantial increase in the size of the goal set suggests that our first condition for goal setting to be an effective tool for orchestrating existing institutions—agreement on a small and manageable set of goals—will not be met. Should the set of Sustainable Development Goals fail to provide clear priorities, effectively orchestrating contributions from existing institutions at the implementation stage will become a very demanding operation.

Moving on to review the 2030 Agenda for Sustainable Development in terms of our second condition—a clearly specified, internally coherent, and hierarchically ordered goal structure—does not significantly boost optimism. The Sustainable Development Goals appear more as a *collection* of about equally important priorities than a hierarchically structured *system*. This is not surprising, as the goals and targets were derived from existing commitments expressed in various international instruments. Therefore, the organization of the Sustainable Development Goals more or less mirrors the fragmented and decentralized structure of international law (Kim 2016). Admittedly, however, some sense of relative importance might be inferred from the order in which the goals appear; in the absence of any other clue, goals listed at the very beginning may plausibly be seen as being more important than those appearing at the end. Goal 1 on eradicating poverty in all its forms and dimensions, including extreme poverty, is indeed noted as "the greatest global challenge" for sustainable development (UNGA 2015, par. 2). One may, perhaps, also argue that "achieving sustainable development in its three dimensions [is the] overarching goal" (UNGA 2012, par. 56). Both of these steps could help, but for the latter to provide firm guidance, "sustainable development in its three dimensions" must be clearly defined. While "our vision" of the 2030 Agenda for Sustainable Development could have elaborated on what that phrase means, it simply reiterates in three paragraphs the key priority areas already embedded in the 17 individual goals (Kim 2016). Furthermore, by reiterating "The Future We Want," the Open Working Group on Sustainable Development Goals considered "poverty eradication, changing unsustainable and promoting sustainable patterns of consumption and production and protecting and managing the natural resource base of economic and social development" as equally important "overarching objectives of and essential requirements for sustainable development" (Open Working Group on Sustainable Development Goals 2014, 6). The negotiators themselves may have a common and fairly precise interpretation of these sentences, but most leaders and staff members of existing institutions whose activities are to be orchestrated will likely not be able to derive clear behavioral implications for their particular activities.

Furthermore, the Sustainable Development Goals are unclear as to what purpose each of the three dimensions—economic, social, and environmental—should serve for sustainable development. For example, Goals 13, 14, and 15, which are commonly identified as environmental goals, are not given a common purpose. Probably the closest to such is the determination of the international community expressed in the preamble of the 2030 Agenda "to protect the planet from degradation ... so that it can support the needs of the present and future generations" (UNGA 2015). This statement recognizes a healthy planetary environment as a prerequisite for meeting the needs of people, but it is obviously not intended as a statement of a goal, and hence lacks the necessary goal attributes such as content and intensity (Kim 2016; see also Latham and Locke 1991). The absence of an overarching environmental goal implies, among other things, the risk of environmental problem shifting. Environmental problem shifting takes place when measures specifically devised for environmental protection result in a transfer of damage or hazards from one area to another or transform one type of pollution into another (Teclaff and Teclaff 1991; see also Kim and van Asselt 2016). For example, replacing gasoline with corn ethanol for the purpose of climate change mitigation may shift net environmental impacts toward increased eutrophication and greater water scarcity (Yang et al. 2012), partially shifting problems from Goal 13 to Goals 6, 14, and 15. As such, the nonhierarchical organization of the Sustainable Development Goals has created an environment conducive to problem *shifting* rather than problem *solving*. It is conceivable that, even in an ideal world where all the goals and targets are met individually, the outcome may not necessarily be the desired state of sustainable development (Kim 2016).

As for the third condition—agents being willing and able to work together to achieve the goals—a fairly safe prediction seems to be that a number of existing institutions will see new opportunities in the Sustainable Development Goals for expanding their own domains or attracting new resources, and therefore offer their services in support of one particular goal or target. Everything else constant, the larger and more diverse the menu of (sub)goals, the more opportunities will arise for such win-win alignments. Many of these alignments will likely facilitate and enhance implementation of specific goals. Some may also foster new partnerships beyond the UN system itself (Andresen and Iguchi, this volume, chapter 8), or "gradually produce a spontaneously emerging division of labor among overlapping institutions" (Gehring and Faude 2014, 471). Where there is a clear overlap, the Sustainable Development Goals may spur "clustering" of related multilateral agreements. For example, Goal 15 could further strengthen the existing cluster around the Liaison Group of Biodiversity-related Conventions by serving as its group goal. In fact, the targets under Goal 15 are traceable to the objectives of the major biodiversity-related treaties.[1] Similarly, Goal 14 reinforces the marine environment treaty cluster around the UN Convention on the Law of the Sea, the implementation of which is specifically mentioned as critical for achieving the goal (target 14.c).

However, the net effect of such developments will hardly amount to a significant reduction in the overall fragmentation of global governance. It will require more than just organizational clustering to effectively address system-wide interactive effects and enhance institutional coherence across earth's subsystems (Walker et al. 2009; Kim and Bosselmann 2013). This is challenging in part due to how the Sustainable Development Goals are themselves designed based on the functionalist thinking that underpins the UN system (Hey 2010). While "siloization" is precisely what the goals are supposed to counteract, they are themselves formulated using a reductionist approach (Bernstein, this volume, chapter 9). In effect, for example, the three environmental goals pertaining to "climate action" (Goal 13), "life below water" (Goal 14), and "life on land" (Goal 15) may reinforce the fragmentation of global environmental governance along three arbitrary spatial domains—that is, the land, the ocean, and the atmosphere (Kim 2016). Having identified themselves as belonging to one of the three, treaty bodies of relevant multilateral agreements or other intergovernmental organizations may lack sufficient incentives to go beyond the conceptual divide and accept orchestration at a higher level.

At a more fundamental level, there is the question of whether the Sustainable Development Goals will be instrumental in striking "a just balance among the economic, social and environmental needs of present and future generations" (Open Working Group on Sustainable Development Goals 2014, par. 9). Here, the High-level Political Forum, as the "orchestrator of orchestrators" (Bernstein, this volume, chapter 9), faces a daunting challenge to address critical trade-offs in the absence of a clear normative vision of how the world would look when all of the Sustainable Development Goals and targets are met (Kim 2016). Most of the goals are framed in terms of *either* human development (Goals 1, 2, 3, 4, 5, 8, 9, 10, 11, 16, and 17) *or* environmental protection (Goals 13, 14, and 15), with relatively few that could be seen as cutting across the two policy domains (Goals 6, 7, and 12). The document is largely silent on how to reconcile the "right to development" and the "rights of nature" with a view to achieving "harmony with nature" (Open Working Group on Sustainable Development Goals 2014, paras. 7 and 9).[2] The failure of the 2030 Agenda for Sustainable Development to address this "normative anarchy" might have been a strategic choice on the part of the states to enable political agreement on the Sustainable Development Goals (Bernstein, this volume, chapter 9).

At least two measures may help orchestrating institutions to not only conform with individual targets but to realize the spirit of the Sustainable Development Goals. First, the High-level Political Forum should clarify the ultimate goal of the goals by, for example, initiating a global dialogue through which an updated definition of sustainable development would eventuate (Kim 2016; see also Muys 2013 and Griggs et al. 2013). The oft-quoted Brundtland version is unfit as the guiding definition because it allows for various interpretations of what sustainable development might mean, especially in relation to integrating the economic, social, and environmental objectives (Kim and Bosselmann 2015). Second, states and other nonstate actors should recognize international law as a normative context within which the Sustainable Development Goals and targets operate (Kim 2016). These goals and targets did not emerge from, and were not inserted into, a normative vacuum. They are grounded in and made consistent with international law. Accepting international law as a normative context would help avoid normative anarchy through the potential application of the principle of integration in cases where the goals and targets conflict (see for example, Voigt 2009).

Conclusion

The Sustainable Development Goals are a demanding encounter between an ambitious and loosely integrated declaration of goals and a complex and fragmented system of global governance. Many existing regimes and organizations will find opportunities for contributing within their own particular domains, and some will also find new opportunities for expanding their domains through collaboration or through penetration of other institutions' domains. Because of their "ecumenical" diversity and soft priorities, the Sustainable Development Goals are, however, not likely to serve as effective instruments for fostering convergence. The 2030 Agenda for Sustainable Development does not provide an overarching goal or norm that can serve as a platform for more specific goals and provide an integrating vision of what long-term *sustainable* development in the Anthropocene means and requires. In the absence of such an overarching principle and vision, the impact of the Sustainable Development Goals on global governance will likely materialize primarily as spurring some further clustering or realignment of existing regimes and organizations within crowded policy domains. These kinds of reforms may in several instances be nontrivial achievements. Yet, given the nature of the challenge (see Young et al., this volume, chapter 3), the Sustainable Development Goals cannot be expected to generate major architectural reforms or new integrative practices that will significantly reduce the fragmentation of the global governance system at large.

Notes

1. For example, the Convention on Biological Diversity (targets 15.1, 15.2, 15.4, 15.5, 15.6, and 15.8), the Ramsar Convention on Wetlands (target 15.1), the UN Convention to Combat Desertification (targets 15.1 and 15.3), the International Treaty on Plant Genetic Resources for Food and Agriculture (target 15.6), and the Convention on International Trade in Endangered Species of Wild Fauna and Flora (target 15.7).

2. The notion of "harmony with nature" referred to in the Outcome Document of the Open Working Group on Sustainable Development Goals (2014) is a value-laden concept, whose meaning has been gradually elaborated in a series of UN General Assembly resolutions and reports of the Secretary-General on Harmony with Nature. See http://www.harmonywithnatureun.org/.

References

Abbott, Kenneth W., Philipp Genschel, Duncan Snidal, and Bernhard Zangl. 2015. Orchestration: Global Governance through Intermediaries. In *International Organizations as Orchestrators*, ed. Kenneth W. Abbott, Philipp Genschel, Duncan Snidal, and Bernhard Zangl, 3–36. Cambridge, UK: Cambridge University Press.

Abbott, Kenneth W., and Duncan Snidal. 2009. Strengthening International Regulation through Transnational New Governance: Overcoming the Orchestration Deficit. *Vanderbilt Journal of Transnational Law* 42: 501–578.

Abbott, Kenneth W., and Duncan Snidal. 2010. International Regulation without International Government: Improving IO Performance through Orchestration. *Review of International Organizations* 5: 315–344.

Allison, Graham T. 1971. *Essence of Decision: Explaining the Cuban Missile Crisis*. 1st ed. Boston: Little Brown and Co.

Biermann, Frank, and Steffen Bauer, eds. 2005. *A World Environment Organization: Solution or Threat for Effective International Environmental Governance?* Aldershot: Ashgate.

Boyle, Alan. 2007. Relationship between International Environmental Law and Other Branches of International Law. In *The Oxford Handbook of International Environmental Law*, ed. Daniel Bodansky, Jutta Brunnée, and Ellen Hey, 125–146. Oxford: Oxford University Press.

Costanza, Robert, Jacqueline McGlade, Hunter Lovins, and Ida Kubiszewski. 2015. An Overarching Goal for the UN Sustainable Development Goals. *Solutions* 5: 13–16.

Danielsen, Finn, Hendrien Beukema, Neil D. Burgess, Faizal Parish, Carsten A. Bruhl, Paul F. Donald, Daniel Murdiyarso, et al. 2008. Biofuel Plantations on Forested Lands: Double Jeopardy for Biodiversity and Climate. *Conservation Biology* 23: 348–358.

Fargione, Joseph, Jason Hill, David Tilman, Stephen Polasky, and Peter Hawthorne. 2008. Land Clearing and the Biofuel Carbon Debt. *Science* 319: 1235–1238.

Gehring, Thomas, and Benjamin Faude. 2014. A Theory of Emerging Order within Institutional Complexes: How Competition among Regulatory International Institutions Leads to Institutional Adaptation and Division of Labor. *Review of International Organizations* 9: 471–498.

Griggs, David, Mark Stafford-Smith, Owen Gaffney, Johan Rockström, Marcus C. Öhman, Priya Shyamsundar, Will Steffen, et al. 2013. Sustainable Development Goals for People and Planet. *Nature* 495: 305–307.

Hey, Ellen. 2010. The MDGs, Archeology, Institutional Fragmentation and International Law: Human Rights, International Environmental and Sustainable (Development) Law. In *Select Proceedings of the European Society of International Law*, Vol. 2, eds. Hélène R. Fabri, Rüdiger Wolfrum, and Jana Gogolin, 488–501. Oxford: Hart Publishing.

International Court of Justice. 1997. Gabčíkovo-Nagymaros Project (Hungary/Slovakia), Judgment. *ICJ Reports* 1997: 7–81.

Khodyakov, Dmitry M. 2007. The Complexity of Trust-Control Relationships in Creative Organizations: Insights from a Qualitative Analysis of a Conductorless Orchestra. *Social Forces* 86: 1–22.

Kim, Rakhyun E. 2016. The Nexus between International Law and the Sustainable Development Goals. *Review of European, Comparative and International Environmental Law* 25 (1): 15–26.

Kim, Rakhyun E., and Klaus Bosselmann. 2013. International Environmental Law in the Anthropocene: Towards a Purposive System of Multilateral Environmental Agreements. *Transnational Environmental Law* 2: 285–309.

Kim, Rakhyun E., and Klaus Bosselmann. 2015. Operationalizing Sustainable Development: Ecological Integrity as a *Grundnorm* of International Law. *Review of European, Comparative and International Environmental Law* 24: 194–208.

Kim, Rakhyun E., and Harro van Asselt. 2016. Global Governance: Problem-shifting in the Anthropocene and the Limits of International Law. In *Research Handbook on International Law and Natural Resources*, ed. Elisa Morgera and Kati Kulovesi, 473–495. Cheltenham: Edward Elgar.

Latham, Gary P., and Edwin A. Locke. 1991. Self-Regulation through Goal Setting. *Organizational Behavior and Human Decision Processes* 50: 212–247.

Muys, Bart. 2013. Sustainable Development within Planetary Boundaries: A Functional Revision of the Definition Based on the Thermodynamics of Complex Social-Ecological Systems. *Challenges in Sustainability* 1: 41–52.

Nye, Joseph S. 2004. *Soft Power: The Means to Success in World Power*. New York: Public Affairs.

Oberthür, Sebastian. 2002. Clustering of Multilateral Environmental Agreements: Potentials and Limitations. *International Environmental Agreement: Politics, Law and Economics* 2: 317–349.

Olson, Mancur. 1982. *The Rise and Decline of Nations: Economic Growth, Stagflation, and Social Rigidities*. New Haven: Yale University Press.

Open Working Group on Sustainable Development Goals. 2014. Proposal of the Open Working Group for Sustainable Development Goals. UN Doc. A 68/970/12.

Pauwelyn, Joost. 2003. *Conflict of Norms in International Law: How WTO Law Relates to Other Rules of International Law*. Cambridge, UK: Cambridge University Press.

Pavoni, Riccardo. 2010. Mutual Supportiveness as a Principle of Interpretation and Law-Making: A Watershed for the "WTO-and-Competing-Regimes" Debate? *European Journal of International Law* 21: 649–679.

Pierson, Paul. 2000. Increasing Returns, Path Dependence, and the Study of Politics. *American Political Science Review* 94 (2): 251–267.

Risse, Thomas. 2000. "Let's Argue!" Communicative Action in World Politics. *International Organization* 54: 1–39.

Rockström, Johan, Will Steffen, Kevin Noone, Åsa Persson, F. Stuart Chapin, III, Eric F. Lambin, Timothy M. Lenton, et al. 2009. A Safe Operating Space for Humanity. *Nature* 461: 472–475.

Sanwal, Mukul. 2004. Trends in Global Environmental Governance: The Emergence of a Mutual Supportiveness Approach to Achieve Sustainable Development. *Global Environmental Politics* 4: 16–22.

Scott, W. Richard. 1981. *Organizations: Rational, Natural, and Open Systems*. Englewood Cliffs: Prentice Hall.

Teclaff, Ludwik A., and Eileen Teclaff. 1991. Transfers of Pollution and the Marine Environment Conventions. *Natural Resources Journal* 31: 187–211.

Underdal, Arild. 2002. Conclusions: Patterns of Regime Effectiveness. In *Environmental Regime Effectiveness: Confronting Theory with Evidence*, Edward L. Miles, Arild Underdal, Steinar Andresen, Jørgen Wettestad, Jon Birger Skjærseth, and Elaine M. Carlin, 433–466. Cambridge, MA: The MIT Press.

UNEP, United Nations Environment Programme. 2013. Embedding the Environment in Sustainable Development Goals. UNEP Post-2015 Discussion Paper 1. Nairobi: United Nations Environment Programme.

UNGA, United Nations General Assembly. 2000. United Nations Millennium Declaration. UN Doc. A/RES/55/2.

UNGA, United Nations General Assembly. 2005. 2005 World Summit Outcome. UN Doc. A/RES/60/1.

UNGA, United Nations General Assembly. 2012. The Future We Want. UN Doc. A/RES/66/288 (Annex).

UNGA, United Nations General Assembly. 2015. Transforming Our World: The 2030 Agenda for Sustainable Development. UN Doc. A/RES/70/1.

UN Framework Convention on Climate Change. 2006. Views on the Paper on Options for Enhanced Cooperation among the Three Rio Conventions Submissions from Parties. UN Doc. FCCC/SBSTA/2006/MISC.4.

Velders, Guus J. M., Stephen O. Andersen, John S. Daniel, David W. Fahey, and Mack McFarland. 2007. The Importance of the Montreal Protocol in Protecting Climate. *Proceedings of the National Academy of Sciences of the United States of America* 104: 4814–4819.

Voigt, Christina. 2009. *Sustainable Development as a Principle of International Law: Resolving Conflicts between Climate Measures and WTO Law*. Leiden: Martinus Nijhoff.

von Moltke, Konrad. 2005. Clustering International Environmental Agreements as an Alternative to a World Environment Organization. In *A World Environment*

Organization: Solution or Threat for Effective International Environmental Governance? ed. Frank Biermann and Steffen Bauer, 175–204. Aldershot: Ashgate.

Walker, Brian, Scott Barrett, Stephen Polasky, Victor Galaz, Carl Folke, Gustav Engström, Frank Ackerman, et al. 2009. Looming Global-Scale Failures and Missing Institutions. *Science* 325: 1345–1346.

Weeramantry, C. G. 2004. *Universalising International Law*. Leiden: Martinus Nijhoff Publishers.

Wolfrum, Rüdiger, and Nele Matz. 2003. *Conflicts in International Environmental Law*. Berlin: Springer.

Yang, Yi, Junghan Bae, Junbeum Kim, and Sangwon Suh. 2012. Replacing Gasoline with Corn Ethanol Results in Significant Environmental Problem-Shifting. *Environmental Science & Technology* 46: 3671–3678.

Zelli, Fariborz, and Harro van Asselt. 2013. The Institutional Fragmentation of Global Environmental Governance: Causes, Consequences, and Responses. *Global Environmental Politics* 13: 1–13.

11 Financing the 2030 Agenda for Sustainable Development

Tancrède Voituriez, Kanako Morita, Thierry Giordano, Noura Bakkour, and Noriko Shimizu

In March 2002, the Monterrey Conference on Financing for Development called for greater mobilization of private finance to support poverty reduction through economic growth and job creation (UNGA 2002). In September of that same year, at the Johannesburg Summit on Sustainable Development, the UN Secretary-General launched what was supposed to boost private-sector involvement in developing countries: the so-called "Type II outcomes" or "partnerships," consisting of a series of commitments and action-oriented coalitions. The follow-up, however, has been quite disappointing (Ramstein 2012). Although evidence of such mobilization has so far been scarce, the Addis Ababa Action Agenda of the Third International Conference on Financing for Development strongly emphasizes the need to mobilize domestic and international private finance for sustainable development (UNGA 2015). In this chapter, we explain why crowding in private finance is now an unavoidable component when designing financing for global development agreements. After looking at investment needs (section 1), new blending finance facilities (section 2), and public-private partnerships (section 3), we highlight the need and opportunities for financial partners to bend their learning curve, particularly with respect to successfully mobilizing private sector investment (section 4). We then examine policy implications and look at possible donor interventions that could help narrow the gap between the high expectations placed on public-private partnerships and what they actually achieve (section 5). We conclude with our key messages.

Unprecedented Investment Needs

It is hardly surprising that integrating the Millennium Development Goals, launched in 2000, into a new and broader set of Sustainable Development Goals will push up the overall cost of achieving these. The Sustainable

Development Goals cover many more topics than the Millennium Development Goals, and they are more ambitious—targeting, for example, "zero poverty" and "zero hunger"—and placed within a universal perspective.

This widening of the development agenda raises specific questions on implementation, particularly in the area of funding. The UN Intergovernmental Committee of Experts on Sustainable Development Financing examined this issue and drew up an inventory of funding needs and sources (UNGA 2014a). This review of the investment needed to tackle climate change and broader sustainable development issues was fine-tuned by the Sustainable Development Solutions Network a few months ahead of the Third International Conference on Financing for Development in Addis Ababa (SDSN 2014).

The UN Intergovernmental Committee of Experts on Sustainable Development Financing has evaluated the financing needs at US$35–195 billion per year to eradicate extreme poverty, and US$5–7 trillion to cover investment needs in infrastructure, with an additional US$2.5–3.5 trillion to develop small and medium enterprises (UNGA 2014a, 10). In the energy sector, the International Energy Agency estimates that US$48 trillion global investment is required to meet projected energy demand between now and 2035. In the low-carbon 2-degree scenario, total investment needs of the energy sector are projected to be US$53 trillion, which is 10% up on the business-as-usual scenario. Across the economy, the Global Commission on Economy and Climate estimates that in a low-carbon scenario, infrastructure requirements would be around US$90 trillion between 2015 and 2030 (New Climate Economy 2014). Interestingly, this is "only" US$4 trillion more than investment needs under the business-as-usual scenario, partly due to spillovers in cost savings across sectors and technologies. While the exact numbers will depend on GDP and population growth rates as well as technological developments, both estimates suggest that investment shifts will need to be substantially larger than incremental investments.

Estimates of the order of magnitude given in the literature and compiled by the Intergovernmental Committee of Experts on Sustainable Development Financing show that the annual requirements are at least 20 times higher than the annual official development assistance, which reached a record level in 2013 of around US$134 billion. This official development assistance will grow only slightly—due to present and future burdens on donor country public finances—and will never match the financing needs in the broadest sense. It is possible that the announcement of the new Sustainable Development Goals will kick-start mobilization—as was the case between 2000 and 2005 following the launch of the Millennium

Development Goals and the implementation of the Heavily Indebted Poor Countries initiative—but it seems unlikely to incite a bifurcation or sea change in the long-term trend of net official development assistance, which follows a very steady trajectory.

The bulk of the additional funds required to cover the financing needs of the 2030 Agenda for Sustainable Development must therefore come from other sources of long-term financing—for example institutional investors such as pension funds, insurance companies, and sovereign wealth funds. The Intergovernmental Committee of Experts on Sustainable Development Financing has stated that public and private savings amount to US$22 trillion, and financial assets US$218 trillion. Reallocating some of these resources would theoretically cover all of the estimated needs (UNGA 2014a, 11). The United Nations estimates that institutional investors alone hold financial assets worth US$75–85 trillion. Pension funds, life insurance companies, and sovereign wealth funds (together holding US$60 trillion in assets) have financial tools (long-term liabilities) that are compatible with the long-term horizon required for some investments in the 2030 Agenda for Sustainable Development (UNGA 2014a). As highlighted in the UN report on the implementation of the Monterrey Consensus and the Doha Declaration (UNGA 2014b), these "long-term investors today do not invest enough in the long-term direct investment necessary for sustainable development, both in developing countries and rich countries—regardless of the institutional and regulatory framework" (UNGA 2014b, 7). Leveraging private investment with a limited amount of official development assistance—or "doing more with less"—is the core principle of blending, which we discuss next.

New Blending Facilities

The definitions of "blending" or "blended finance" differ across institutions (UNGA 2014a). One approach distinguishes between sources of financing on the basis of their institutional nature: blending thus corresponds to a mixture of public and private funding. This approach, however, is often misleading. Public funding is not channeled through grants alone. For instance, part of France's official development assistance is financed through loans, which are themselves refinanced through private savings on capital markets. Equally, private financing does not always involve loans, as is the case of grants from the Bill and Melinda Gates Foundation, which have nonprofit goals and therefore differ little from official development assistance.

The second approach emphasizes the distinction between the different types of financing instruments: loans, grants, guarantees, and equity investments are combined within the same operation. This makes it possible to establish a blending typology and to identify emerging innovations more precisely, along with their scope and potential application.

In this respect, blending most traditionally involves combining a variety of instruments—basically loans and grants—from a single institution. This type of blending translates into subsidized loans and represents the core business of development finance institutions such as the European Investment Bank, the French Development Agency, and the German development bank KfW.

The second type of blending involves combining funding from financial and nonfinancial partners. This is the model used by the Global Environment Facility, as well as by the eight blending facilities launched by the European Commission in 2007. The EU approach to blending involves using targeted EU grants to mobilize nongrant funding under the lead of a European multilateral (for example, the European Investment Bank) or national finance institution (for example, the French Development Agency or the German KfW). EU blending comprises direct investment (41%); interest rate subsidy grants (19%); technical assistance (32%); risk capital (4%); and guarantees (3%). Average EU grant size lies in the range of €5–10 million. There are eight EU global facilities involving blending. The sectors covered are mostly energy (35%), transport (26%), water (20%), followed by support for small and medium enterprises (11%) and social (5%) and information and communication technologies (3%). Leveraged resources figures for the EU blending facilities since 2007 show that €1.6 billion of EU grants unlocked €42 billion of additional financing (grants, loans, and investments). Climate change windows in EU blending facilities were announced in 2010. They provide transparent tracking of all climate change–related projects in the EU regional blending facilities, as well as the opportunity to draw additional resources for climate change adaptation and mitigation projects. More than €700 million are targeted as "climate finance" in EU blending (European Commission 2013).

In its "Beyond 2015: Toward a Comprehensive and Integrated Approach to Financing Poverty Eradication and Sustainable Development," released in July 2013, and in its 2012 communication (European Commission 2012), the European Commission proposed a future development financing framework that reinforces the links between public and private finance and domestic and international resources. The communication argues that private finance is the "key driver of growth" and that countries should "use public resources to invest in areas that leverage private investments towards policy priorities" (European Commission 2013). This explicit linking of the use of public resources to leverage private investment is not a new theme for the commission, as it stands as a key feature of its Agenda for Change policy.

In recent years, most blending operations have provided subsidized loans to the public sector (some 90% of recipient projects target public investment) in developing countries. For the EU 2014–2020 budgetary period, the intention is much more explicitly to use EU aid to subsidize or incentivize private-sector loans—which is tantamount to an admission that this was not previously the case. In the 2013 communication, the European Commission thus argued that the "blending of grants with loans and equity, as well as guarantee and risk-sharing mechanisms can catalyze private and public investments, and the European Union is actively pursuing this" (European Commission 2013). While the vast majority of existing blending operations have been in support of the public sector, the European Commission's plans for the future include a significant scale-up of blended finance for the private sector.

According to the European Commission, the €1.2 billion grant contribution from the EU budget, the European Development Fund, and member governments have leveraged loans of development finance institutions worth €32 billion, unlocking project financing of at least €45 billion, in line with EU policy objectives (European Commission 2013). Two points should be underlined here. Even though development finance institutions borrow on international capital markets to finance their loan, the loan-grant blending facility is a public-public partnership. Secondly, as Bilal and Krätke observe, "estimates on the amount of funding invested in and leveraged through blending facilities vary considerably. The European Investment Bank has noted leverage ratios of 8 times the EU-budget contribution, whereas the European Commission has noted leverage ratios of up to 31 times. Measures of leverage are also confused, notably confusing the grant-to-loan-component ratio on the one hand with the grant-to-total-cost ratio on the other" (Bilal and Krätke 2013, 2).

The European Commission's position on blending contrasts with the more cautious, evidence-based approach promoted by the European Parliament, whose June 2013 Resolution on Financing for Development called on the European Union "to properly evaluate the mechanism of blending loans and grants—particularly in terms of development and financial additionality, transparency and accountability, local ownership and debt risk—before continuing to develop blending loans and grants" (European

Commission 2013). In their study "Financing for Development Post-2015: Improving the Contribution of Private Finance," commissioned by the European Parliament's Committee on Development, Griffiths, Martin, Pereira, and Strawson stress that "policy makers seeking to maximise the role that private finance can play in development must recognise three key limitations" (2014).

First, the authors write, private finance flows predominantly toward middle- and high-income countries—a point flagged already in 1990 by Lucas (1990). Second, in developing countries the private sector is dominated by micro, small, and medium enterprises that find it particularly difficult to access external private financing sources. Close to 80% of these operate in the informal economy, which not only reduces the government's tax base and impacts decent working practices, but also constitutes a major obstacle for both enterprises and individuals to access finance, insurance, social safety nets, and formal-sector business opportunities. Finally, the incentives for the private sector to invest in the protection and provision of public goods are limited, since these are by definition non-rivalrous and non-excludable. The World Bank estimates that over the last decade 80–85% of all infrastructure investments in developing countries have been funded by the public sector (World Bank 2014).

Revisiting the Case of Public-Private Partnerships

Public-private partnerships might be a solution in this situation. A public-private partnership is a contract drawn up between public and private entities to mitigate the level of risk-taking for all parties and create a win-win situation: The public entity can trigger the most needed investments despite an eventual lack of funding; the private entity can expand its business by building and operating the required infrastructure while benefiting from more effective risk-sharing. It could be tempting to replicate these public-private partnerships in poor countries for the obvious reason that public finance, although scarce, remains the main source of funding and that it would be more efficient to use these funds to leverage private finance instead of replacing it. Yet is this really a viable option, given that most investment in poor countries is characterized by its lumpiness, on the one hand, and the long maturity and low liquidity of most assets on the other?

A first strand of the literature attempts to craft the main features of public-private partnerships into a feasible mechanism for developing countries. Yet, so far, project evaluations show that successes have been few and far between, both in terms of the amount of private funding raised and the projects' impacts on growth, employment, and poverty alleviation (Griffiths, Martin, Pereira, and Strawson 2014). These findings have opened avenues to explore new or unconventional public-private partnerships in very different economic and social sectors (Chattopadhay and Batista 2014; Hossain and Ahmed 2014). However, these more innovative public-private partnerships do not seem as yet to attain the scale and role played by public-private partnerships in developed or emerging countries.

Drawing lessons on past experiences, a second strand focuses on the conditions required to develop public-private partnerships and ensure they deliver their intended economic and social outcomes (Trebilcock and Rosenstock 2015). One drawback is that most past experiences are from emerging and developed countries. As a result, new outcome-based mechanisms have been developed to reinforce the social impacts of public-private partnerships and balance the risk/profit trade-offs for private investors (for example, development impact bonds). While very promising, these improvements are of little direct help to poor countries. Although they restore the development impact of public-private partnerships, they fail to reinforce their applicability to poor countries, given that the credit risk for the private sector is still excessively high.

A third strand of the literature points out the possibility of blending public and private finance in different ways, driven by the development of additional, innovative sources of finance. For example, over the past few years, many developing countries have accessed international capital markets for the first time. Government debt thus constitutes a "new" source of public finance, offering the possibility of developing innovative blending mechanisms that combine government debt, donor facilities (grants or concessional loans), and private finance. The question is whether blended finance can have a strong enough impact on public and private risk- and responsibility-sharing to incentivize debt holders, equity holders, governments, and guarantee providers to agree on new modalities for doing business in poor countries.

In their analysis of how government intervention has evolved against the backdrop of the financial crisis and the related constraints on the supply of capital, Hellowell, Vecchi, and Caselli (2015, 74) identify five categories of state intervention recently used to stimulate public-private partnerships. Since developing countries face exactly the same constraints on a permanent basis, not only during crises, it would be useful to assess the extent to which blended finance actually matches these same solutions.

The first category relates to projects where the private operator's revenue depends on user payments. The state entity commits to make scheduled debt repayments to lenders should the private operator default. This means that lenders bear only the construction risks. In poor countries, this kind of state guarantee is undermined by low levels of public resources, weak governance, and high political risks. As the state is ill-equipped to act as the insurer of last resort, this responsibility must necessarily be assumed by another actor.

On very similar lines, the second option also deals with the mitigation of credit risks, but instead of relying on a direct state guarantee, capital is set aside beforehand to provide an indirect credit guarantee through subordinated debt. This debt is only used by the private operator in the event that he has difficulty in paying off the senior debt. The state entity is endowed with sufficient capital to significantly improve credit quality. This mechanism aims at crowding in long-term investors. However, state entities in many developing countries have been poorly managed, and their credit rating is no better than that of the state itself, which of course detracts from the reliability of this type of mechanism. This means that an external actor with a good credit rating will need to provide this subordinated debt in order to ensure credit quality.

A third category of intervention involves changing a project's capital structure. This requires the private operator to increase its equity share in the project so as to limit its debt exposure and strengthen its capacity to cope with cash-flow fluctuations. This option might be limited in poor countries, where raising equity capital is already a hugely difficult challenge. Alternatively, a higher share of government equity capital would lower the credit risk. Government equity is supposed to increase the transparency of the contract, but poor governance in most countries makes government equity unlikely to enhance contract transparency; quite the reverse.

A fourth option targets the refinancing of the loan at maturity. Providing a refinancing guarantee makes it possible to have shorter tenors. It makes the initial project more affordable, a serious advantage in developing countries where the capacity to pay is often rather low. However, the same cautiousness as before applies here, since a developing state can hardly act as a guarantor.

The last option involves a very different mechanism: substituting public debt capital for private finance through a public entity. This entity, most often a state infrastructure bank, becomes the lender to the private operator, with guarantees from insurers and banks. Rather than playing on loan terms and conditions, the purpose here is to overcome initial liquidity constraints to make sure that the project actually takes off. Financing is provided at market rates, as the state bank needs sufficient returns on investment to operate its business. In developing countries, this type of public entity would most likely benefit from state guarantees rather than insurance or bank guarantees, but this would deprive the mechanism of its risk-sharing advantage, as the state would be the sole risk-bearer.

The above options fit nicely into the context of developed economies, but replicating them in poor economies is difficult to envisage, chiefly because the state is required to play a central role in providing capital or guarantees. In either case, developing countries are too hard-pressed to assume such a role due to the lack, and sometimes misuse, of public funds. If public-private partnerships are to gain ground, other actors must take over.

Bending the Learning Curve

How can resources, then, be better mobilized to narrow the funding gap? While many potential new sources of capital and expertise remain untapped (insurers, pension funds, and so forth), innovative financing for the development market is still very conservative, with roughly two-thirds of initiatives based on conventional bonds and guarantees (Dalberg 2014). Should we be scaling and replicating these "tried and tested" instruments rather than designing bespoke pilot solutions? The cost of gathering on-the-ground results and historical performance data means that investors face uncertain and often complex risk profiles, and therefore demand high returns (Dalberg 2014). Here, the well-known effect of endogenous risk appears to kick in: The fewer the track records and metrics, the higher the perceived risk, which naturally discourages new initiatives and further reduces the number of available track records and metrics, and so on. Breaking out of this vicious circle necessarily means changing how the actors perceive the risks involved by deepening their understanding of how this risk can be shared or transferred to maximize economic and non-economic returns—which all entails substantial transaction and research costs.

Donors, international financial institutions, development finance institutions, and philanthropists have all developed initiatives to entice private operators into working in poor countries where the needs are huge but business conditions uninviting.

Donors' interventions to support public-private partnerships can be clustered into three broad categories. The first relates to enhancing the public finance capacities of poor states, and indirectly to improving their business environment. A wide range of measures can be included in this category, all pertaining to conventional official development assistance operations. Some are crucial to enable private actors to operate: building and strengthening institutions; laws and regulations conducive to private initiative; supporting social (such as education, health, and sanitation) and economic (such as energy, transport, and banking) services provision; and incentivizing the development of a transparent, stable, and predictable investment climate. Unfortunately, the transformative effects of such measures are still too slow and weak to have had significant impact on private investment. As for foreign direct investment, this has indeed increased over recent decades, but with many disparities between sectors and countries. Some suggest that this "old official development assistance model" is dead and unable to deliver, calling on donors to change their modus operandi and focus instead on supporting public-private partnerships (Simon, Schellekens, and de Groot 2014).

A second set of measures targets public-private partnerships directly. It seeks to facilitate the intervention of private operators through the provision of guarantees and insurance. Despite many experiences in this field, the level of investment remains low and reveals the donors' lack of appetite for risk—even though this should be their core business—as more commitments would probably imply more risks and more failures. The third set of measures focuses on the outcomes of private interventions and on what boils down to development impact investment (Dalberg 2014).

The 2014 Dalberg report emphasizes that the focus of innovative financing (blending and public-private partnerships included) is shifting from the mobilization of resources through innovative fundraising approaches to the delivery of positive social and environmental outcomes through market-based instruments. It anticipates three primary drivers of growth in the innovative financing sector, which could all bend the learning curve and narrow the gap between expectations from blending and public-private partnerships and their achievements:

i) Increased use of established financial instruments. These instruments, such as green bonds, which investors can evaluate through existing risk frameworks, will attract new participants, including pension funds and institutional investors. Channeling the proceeds of these instruments to productive development goals will require new standards that specify how funds can be used most effectively.

ii) Expansion into new markets through growth of replicable products. Over the past 10 years, the international development community has experimented with new instruments, such as performance-based contracts. These instruments do not yet have the track record to attract institutional investors, but offer promising opportunities to improve development outcomes in new sectors.
iii) Creation of new innovative financing products. We are seeing the emergence of new products that are promising on paper but have not yet demonstrated results. While these products will represent a small fraction of the market in the short term, Dalberg encourages donor governments and other funders to continue experimenting with them so they can mature into the next important asset class (Dalberg 2014).

As Dalberg (2014) contends, blending has brighter perspectives today than in the past and, as a result, may well enjoy fresh momentum. What still needs to be set up is a carefully designed public policy framework enabling blending and public-private partnerships to meet expectations and enhance the implementation of Sustainable Development Goals. We address this issue in the following section.

Policy Implications

Skeptics will doubtless point out that UN files are already crammed full of texts, treaties, and conventions that, if taken as a whole, cover the three dimensions of sustainable development—economic, social, and environmental. Indeed, it can be legitimately asked what added value the Sustainable Development Goals can bring to this maze of texts.

"If we are serious about implementation, then the bulk of the work will have to be done back at home," comments Csaba Kőrösi, the vice-chair of the Open Working Group on Sustainable Development Goals. He further adds:

> Even though the General Assembly has adopted the Open Working Group's report with its goals and targets, ensuring that it will become a vital part of future negotiations, this in itself will not generate a movement of capital and knowledge. Only national and local plans and projects can achieve this redirection of funds. Banks and institutions will not finance the Sustainable Development Goals; finances and other implementation means will be targeted at actual, tangible projects. On this aspect, there is still much work to be done. Sustainable Development Goals are in place, but most countries do not have national plans and there is certainly a lack of projects. (Kőrösi 2015)

Between global talks at UN headquarters and direct online consultations of world citizens, as in the MyWorld initiative, there is a missing middle: country-level initiatives and appropriation.

One way to foster this national appropriation and bridge the sustainable development implementation gap would be to develop different forward-looking scenarios of potential development paths for 2030 at national and regional levels. These forward-looking exercises are ongoing in some countries with respect to climate change and energy and agriculture.[1] They could be generalized to other Sustainable Development Goals[2] and lead to the formalization of what would be equivalent to intended nationally determined contributions applied to a dashboard of country-relevant Sustainable Development Goals.

On the funding side, what will make (sustainable) development finance transformative? Talks on development and climate financing strongly emphasize the need to mobilize new or additional resources to bridge the gaps in the estimated trillion-dollar figures required to attain the Sustainable Development Goals. It is implicitly assumed that the transformation will stem from "more" money to start with, to be raised by leveraging private finance through public funding through blending and public-private partnership mechanisms. There seems to be a consensus among practitioners that if co-benefits and trade-offs between climate and development priorities are managed together, the effectiveness of both climate and development finance would improve, and the trade-offs would fall into line with the transformative ambition of the Sustainable Development Goals. One possible storyline would be that development finance and the blending and public-private partnership mechanisms discussed in this chapter should be simultaneously climate-proofed and scaled up. Certainly, a critical step forward would be to design national investment plans as an equivalent of the intended nationally determined contributions, as they are known from climate governance.

Lastly, it is very likely that Sustainable Development Goals and countries will not receive their fair share of the capital available worldwide. Hence the need to place greater emphasis on the access to and dynamics of financing, and particularly on the question of who is the final payer: the taxpayer or the consumer? The taxpayer from the North or the South? Is it the rich consumer or the poor consumer? Funding for Sustainable Development Goals in the long term is comparable to the issuance of a debt for which the underwriters and the schedule must be defined at the outset. Otherwise, the financing package will not be *sustainable*, neither environmentally nor financially.

Conclusion

Financing for development and climate financing negotiations place great emphasis on the mobilization of new and additional resources to bridge the gaps with the trillion-dollar estimated needs to achieve the Sustainable Development Goals. We have tried to explain in this chapter why blending and public-private partnerships continue to have a high profile among development finance institutions in spite of their limited track records. First, the magnitude of the financing needs for Sustainable Development Goals calls into question the role of official development assistance and requires unprecedented financing flows. Second, development finance institutions have been exploring new blending facilities over the last eight years, switching their focus from new financing instruments to risk sharing among new partners. The diversity of financing needs for Sustainable Development Goals questions the metrics and value of perceived risk and calls for further experimentation—which blending and public-private partnerships can offer. Last, bending the learning curve with scalable multi-partner pilot projects paves the way to more blending and public-private partnerships. It is worth recalling, however, that Sustainable Development Goals and countries will not receive fair shares of the capital available worldwide. Evidence suggests that blending and public-private partnerships could increase the inequality of global capital allocation across countries and sectors if historical trends and risk-sharing practices continue unchanged. Our suggestion that countries should establish intended nationally determined contributions would provide a way of reducing risk for private investors and bringing more fairness into the allocation of blended public and private flows.

Notes

1. See the Deep Decarbonization Pathways Project at http://unsdsn.org/what-we-do/deep-decarbonization-pathways/.

2. See http://unsdsn.org/news/2015/03/13/the-world-in-2050-pathways-towards-a-sustainable-future/.

References

Bilal, Sanoussi, and Florian Krätke. 2013. Blending Loans and Grants for Development: An Effective Mix for the EU? Briefing Note 55. European Centre for Development Policy Management.

Chattopadhay, Tavo, and Olavo Nogueira Batista. 2014. Public-private Partnership in Education: A Promising Model from Brazil. *Journal of International Development* 26 (6): 875–886.

Dalberg. 2014. Innovative Financing for Development. Scalable Business Models that Produce Economic, Social, and Environmental Outcomes.

Economy, New Climate. 2014. *Better Growth. Better Climate.* The Global Commission on Economy and Climate.

European Commission. 2012. Improving EU support to developing countries in mobilising Financing for Development. Recommendations based on the 2012 EU Accountability Report on Financing for Development. Com/2012/0366/Final.

European Commission. 2013. Beyond 2015: Towards a Comprehensive and Integrated Approach to Financing Poverty Eradication and Sustainable Development. Com/2013/531/Final.

Griffiths, Jesse, Matthew Martin, Javier Pereira, and Tim Strawson. 2014. Financing for Development post-2015: Improving the Contribution of Private Finance. European Parliament, Directorate-General for External Policies.

Hellowell, Mark, Veronica Vecchi, and Stefano Caselli. 2015. Return of the State? An Appraisal of Policies to Enhance Access to Credit for Infrastructure-based PPPs. *Public Money & Management* 35 (1): 71–78.

Hossain, Khandker Zakir, and Shafiul Azam Ahmed. 2014. Non-conventional Public-private Partnerships for Water Supply to Urban Slums. *Urban Water Journal* 12 (7): 570–580.

Kőrösi, Csaba. 2015. Negotiating a Common Future: What We Have Learnt from the SDGs. In *Building the Future We Want*, ed. Rajendra Kumar Pachauri, Anne Paugam, Teresa Ribera, and Laurence Tubiana, 74–82. New Delhi: TERI Press.

Lucas, Robert. 1990. Why Doesn't Capital Flow from Rich to Poor Countries? *American Economic Review* 80 (2): 92–96.

Ramstein, Céline. 2012. Rio+20 Voluntary Commitments: Delivering Promises on Sustainable Development? Working Paper 23/12. Institut du Développement Durable et des Relations Internationales, Paris.

Simon, John, Onno Schellekens, and Arie de Groot. 2014. Public-private Partnerships and Development from the Bottom Up: From Failing to Scaling. *Global Policy* 5 (1): 121–126.

Sustainable Development Solution Network. 2014. Financing for Sustainable Development: Implementing the SDGs through Effective Investment Strategies and Partnerships.

Trebilcock, Michael, and Michael Rosenstock. 2015. Infrastructure Public-private Partnerships in the Developing World: Lessons from Recent Experience. *Journal of Development Studies* 51 (4): 335–354.

UNGA, United Nations General Assembly. 2002. Outcome of the International Conference on Financing for Development, Monterrey, Mexico, March 18–22, 2002. Report of the Secretary-General. UN Doc. A/57/344.

UNGA, United Nations General Assembly. 2014a. Report of the Intergovernmental Committee of Experts on Sustainable Development Financing. UN Doc. A/69/315.

UNGA, United Nations General Assembly. 2014b. Follow-up to and Implementation of the Monterrey Consensus and Doha Declaration on Financing for Development. UN Doc. A/69/358.

UNGA, United Nations General Assembly. 2015. Addis Ababa Action Agenda of the Third International Conference on Financing for Development. UN Doc. A/RES/60/313.

World Bank. 2014. Overcoming Constraints to the Financing of Infrastructure. Report prepared by the Staff of the World Bank Group for the G20 Investment and Infrastructure Working Group, February 2014. Washington, DC: The World Bank.

12 Toward a Multi-level Action Framework for Sustainable Development Goals

Joyeeta Gupta and Måns Nilsson

The Sustainable Development Goals in the 2030 Agenda for Sustainable Development (UNGA 2015) go beyond the earlier Millennium Development Goals. They do not simply focus on taking action in the developing world supported by development cooperation agencies and development banks (Bello 2013; Sanwal 2012). Instead, the Sustainable Development Goals are meant to be universally relevant and aim at directing attention at developmental and environmental issues globally, "while taking into account different national realities, capacities and levels of development and respecting national policies and priorities" (UN 2012).

The process of developing Sustainable Development Goals has been a hybrid of four different approaches. First, inspired by the process of the Millennium Development Goals, the Sustainable Development Goals reflect a top-down approach being centrally negotiated and adopted at the UN level. Second, the process has been informed and influenced by epistemic communities such as the Sustainable Development Solutions Network and the International Council for Science. The goals also focus not just on halving the number of poor people or of people without access to basic resources, but at ending all poverty everywhere, thus reacting to the human rights critique on the Millennium Development Goals (Alston 2005; Redondo 2009; Robinson 2010; Dorsey et al. 2010). Third, at the same time the Sustainable Development Goals have also responded to social movements, as represented by civil society groups and nongovernmental organizations that have promoted the articulation of goals that then became adopted at global level. Fourth, many UN agencies have tried to solicit ideas about goals through a bottom-up process. The result of this hybrid approach has been a framework for Sustainable Development Goals that aspires to reflect the priorities and values of people and governments around the world.

The 2030 Agenda for Sustainable Development (UNGA 2015) is an attempt to inspire, shape, and direct policies and implementation on the ground. This is not the first time such an attempt has been made. A similar effort was undertaken in Agenda 21, adopted at the 1992 UN Conference on Environment and Development (Rio Declaration 1992). However, while Agenda 21 also listed targets and timetables and unleashed activities at multiple levels of governance through a multiplicity of actors, a wide range of the targets were not actually achieved, and over time the interest in Agenda 21 has faded.

Although many actors can adopt goals and promote their adoption at many different levels, the adoption of goals at the global level by the UN General Assembly gives it an authority and democratic legitimacy that no other body within the global arena can aspire to.

Still, if it is to have any real significance for development, a goal framework for development needs to be adopted and owned by those that work on development interventions, including national and subnational state authorities. In other words, it would be necessary for these authorities to also (i) adopt and interpret these goals in their specific context, (ii) to mainstream them in their policy processes, (iii) identify the instruments and mechanisms through which these goals are to be promoted and implemented, and (iv) set up a monitoring and compliance system to report on progress across levels.

The question now is whether—and through what mechanisms—the Sustainable Development Goals can actually induce countries and communities at different levels to engage and take action, and then sustain it over time. This chapter charts out such action mechanisms required to deliver on the Sustainable Development Goals. We address two key questions: First, what main principles for action should be observed for Sustainable Development Goals? Second, how can action mechanisms be structured to enhance coherence in the pursuit of Sustainable Development Goals by different actors? We address these questions through building on examples from the water domain.

Key Principles

This section elaborates on three key principles necessary for actions related to the Sustainable Development Goals to become effective: First, that such actions are approached as a multilevel governance challenge; second, that such actions must address context and level-specific drivers of the problem in question; and, finally, that such actions need to be horizontally and vertically coherent and that they "add up."

Action Principle 1: Govern Sustainable Development Goals across Levels and Actors

The Sustainable Development Goals often reflect global through local public goods (for example, the rule of law in Goal 16, poverty abatement in Goal 1, climate change in Goal 13). Such public goods have the characteristics of being non-rival in nature (one person's use does not necessarily decrease the use by another) and non-exclusive (no one can be excluded from the use of this good). Public goods cannot be easily provided by the market and require state intervention for their provisioning (Kaul et al. 2003; Went 2010; PBL 2011; UNDP 1999; UNIDO 2008). This implies, first, that states need to mainstream the Sustainable Development Goals into their policy and planning processes, and that they can be held accountable for the implementation of the Sustainable Development Goals. However, as has been witnessed in the UN Framework Convention on Climate Change (1992) and its Kyoto Protocol (1997), states can opt in and out of their global responsibilities. Reconciling the challenge of adopting a common global framework for action that is to be implemented by all in such a global governance setting is a key challenge.

At the same time, we are witnessing a rebalancing between centralized governmental steering and decentralized governance in development. A variety of actors and stakeholders now engage in development processes. Indeed, there has been a strong expectation in the process leading toward the Sustainable Development Goals that nonstate actors—whether businesses, nongovernmental organizations, religious groups, educational institutions, or social movements—would adopt the goals and contribute to implementing them within their own realms of action. Paragraph 41 (UNGA 2015) emphasizes "the role of the diverse private sector, ranging from micro-enterprises to cooperatives to multinationals, and that of civil society organizations and philanthropic organizations in the implementation of the new Agenda." In other words, implementation of Sustainable Development Goals must be undertaken by all actors in society, and there has to be a process to ensure ownership by these actors.

At the same time, the very multiplicity of actors and the implied non-hierarchy of one actor over another might lead to situations where different actors have different goals and interests, and these may lead to actions that may counter the overarching goals (Stiglitz 2000). Each of these different actors and stakeholders may decide to prioritize or interpret goals in vastly different and possibly inconsistent ways as they try to align them with their own interests.

This rebalancing from government to governance is supported by both neoliberals who want small government and deeper democracy groups who want to enhance the legitimacy of decision making. In contrast to government, governance is a more elastic term (Doornbos 2001, 95), defined as "a continuing process through which conflicting or diverse interests may be accommodated and cooperative action may be taken" (Commission on Global Governance 1995, 2). It encompasses policy, but also things like self-organizing (Rhodes 1996) and complex patterns of purposive behavior (Rosenau 1992, 4), aiming at mediating differences between people, promoting human rights, and managing resources for social and economic development. In other words, governance is the exercise of authority based on the social distribution of power, through the development and implementation of rules to manage resources for society.

A multilevel governance approach means that the central state does not monopolize action, but interacts in many different ways with actors at subnational, supranational, and international levels (Marks, Hooghe, and Blank 1996, 346; Hooghe and Marks 2003). Governance diffusion and dispersion allows for the internalization of externalities, gives space for heterogeneity and preferences, and allows for multiple arenas of innovation and experimentation (Hooghe and Marks 2003, 235). In terms of the Sustainable Development Goals (Young et al. 2014), this would imply, third, that it would not be enough for just the central state to adopt the Sustainable Development Goals and related targets, but that it would require subnational governments to adopt similar goals and targets while making them contextually and jurisdictionally relevant. This would also provide the context for subnational governance actors to see how they could contribute to goal and target achievement.

This implies that action on Sustainable Development Goals needs to be a balance between top-down and bottom-up approaches (Gupta and Pahl-Wostl 2013) and that nonstate action takes place in the shadow of hierarchy (Scharpf Fritz 1997, 202).

Action Principle 2: Target Drivers of Change at the Appropriate Scale

Environmental and developmental challenges manifest simultaneously from global to local levels. Similarly, drivers and barriers that affect goal achievement exist at multiple levels: household, local, provincial, national, regional, and global. It is always important to understand such drivers before designing action mechanisms to deal with a problem because a local action will be unable to deal with global drivers, and global action may be irrelevant or too blunt for dealing with local drivers.

In an increasingly globalizing world with porous borders, most economic, social, and environmental problems along with human vulnerability, and their drivers, tend to be multilevel in nature (UNDP 2014; UNEP Global Environment Outlook 2013). Some issues are global because they are part of a global cycle or system (for example, climate change or the hydrological cycle), and some accumulate into global challenges (for example, the number of dams worldwide lead to a global environmental challenge; or the number of people without access to sanitation services amounts to a global humanitarian problem).

In some cases, the proximate driving forces and underlying root causes of issues are global (for example, global market flows and the free trade paradigm, or global demographic patterns and the rise of urbanization) and call for global collaboration to address these drivers. At the same time, most of these issues are simultaneously very local—because they call for changes in human behavior at the local level or because they have a locally distinct impact. It is vital that actions to address these problems thus seek to specifically address the driver that operates at each of the distinct levels.

Taking Goal 6 on water and sanitation as an example, the global freshwater system is stressed through over-extraction, over-pollution, and infrastructural efforts that redesign water flows (Vörösmarty et al. 2010; Rockström et al. 2009). The direct drivers affecting water at local and regional levels include infrastructure (for example, dams, supply and sanitation services, irrigation channels, interbasin transfers), large- and small-scale land-use change, and pollution. Indirect drivers include policies and market action in other sectors of society (for example, agriculture, energy, industry, services), climate change and variability, and demographic drivers (Gupta and Pahl-Wostl 2013). For example, energy policy to promote biofuel production can lead to large-scale land-use change from forest area to biofuel production, which then requires water and thus influences the water cycle through both the effects on land-use change and through the additional demand for water for biofuel products (Nilsson and Persson 2012).

Different framings of the water issue underpin the drivers of the water challenge. For example, there is a dominant framing of water as an economic good (Dublin Declaration 1992) and a more recent framing of water as a human right (UNGA 2010). The notion of water as a heritage (European Commission 2000) or as providing ecosystem services (Chopra et al. 2005) provides a different framing of the water issue. These different framings may counter local framings and perceptions of water as a sacred commodity or a gift of God.

The framings shape the types of goals and instruments used to manage water. They underpin direct drivers at the global level, such as global investment and trade patterns, and these may thus impact on water quality and quantity. Rising production and consumption demands to meet the rising standard of life of a growing population may also be a direct driver of water use, as revealed by virtual water trade analyses.

Sometimes, a series of local drivers may influence water policy. For example, one reason the Millennium Development Goal on sanitation fared worse than that on water access was because sanitation is contextually often seen as a taboo, has the alternative of open defecation, and people often do not have the resources or are not willing to pay for sanitation services in relation to other priorities; this may make sanitation an enforced right and may affect the success of policies with respect to sanitation (Obani and Gupta 2014).

A key problem here is the politics of scale. Some countries and actors seek to globalize issues, and others to nationalize them, for a range of political reasons (Gupta 2008 and 2014a). We, however, argue that the achievement of the Sustainable Development Goals requires an understanding of the driving forces that prevent their achievement, so that actions focus simultaneously on these driving forces at the multiple levels on which they operate.

Action Principle 3: Ensure Horizontal and Vertical Coherence in Actions

The framework for Sustainable Development Goals addresses the poverty and human development objectives aiming at meeting basic human needs and promoting development opportunities for all, as well as the ability of the earth system to provide the resources and ecosystem services to sustain this development.

This implies, first, that the actions taken must add up (Weitz, Nilsson, and Davis 2014; Bernstein et al. 2014). In other words, actions must be coherent across levels—from the local level where real action will take place, up to the regional or global levels, where development efforts sometimes need to be coordinated, where monitoring is aggregated, and where finite natural resources need to be managed and used at sustainable levels.

However, there are also other dimensions of coherence that must be observed in the action framework. A second form of coherence refers to coherence in the chain of governance and policy; that is, between policy objectives, instruments, and implementation mechanisms (Nilsson et al. 2012). In the 2030 Agenda for Sustainable Development (UNGA 2015), this implies that the means of implementation (such as capacity building or technology) and nationally determined action plans and institutional arrangements must be coherent with each other and with the agreed goals and targets.

Third, coherence can be examined in the horizontal dimension: to ensure that goals, principles, policies, strategies, and actions within one domain are not in conflict with each other or with those of another domain. This issue has surfaced, since there is an increasing recognition that Sustainable Development Goals will necessarily affect each other (Weitz, Nilsson, and Davis 2014). There are indeed many possible interactions between water and other goals. Some of them are potentially *conflicting*. For example, meeting Goal 7 on access to energy might come at the cost of greater thermal, radiative, or other forms of water pollution, or might lead to displacement and negative livelihood impacts, as in the case of dams. Incentives to produce biofuels such as sugar cane can deplete water tables in water-stressed areas. Extraction of cooling water for thermal power production from rivers may conflict with water requirements for irrigation in agriculture. Trade-offs in resource use and allocations need to be made at the local and subnational level, involving appropriate stakeholder consultations. However, many times, the interactions are *synergetic* (mutually reinforcing). For example, enhanced levels of energy access will enable water access through pumping to households or communities.

This pursuit of horizontal coherence has been popularized in several areas, and in the environmental policy field, the term "integration" has widespread usage, such as in integrated pesticide management, integrated solid waste management, and integrated water resource management. Within development policy, the term "mainstreaming" is more prevalent. Gender mainstreaming calls for gender-sensitive approaches in all sectors and policies of society (Walby 2005, 321; Council of Europe 1998; ECOSOC 2006). Disaster mainstreaming became a critical issue with the International Decade for Natural Disaster Reduction of the 1990s. It called on societies to not see disasters as single, unpredictable events, but to instead see them as a repetitive part of the development strategies of modern society and to call for structural action to prevent and deal with disasters by integrating risk management into policy processes (Benson and Twigg 2007).

A fourth (horizontal) coherence requirement is mainstreaming the Sustainable Development Goals into existing policy arenas. It implies the inclusion of the specific goal into national and subnational governmental and governance plans, strategies, and policies. This implies that such policies need to be (re-)designed, (re-)organized, and evaluated from the perspective of the Sustainable Development Goals—as far upstream as possible (Gupta 2010). A key risk of mainstreaming is that it may not be implemented in fact, and instead results in dilution of policy priorities (Liberatore 1997). This calls for ensuring that the Sustainable Development Goals are accompanied by monitoring and enforcement instruments to ensure actual implementation.

Coherence issues come to light in water policy from global to local levels. Whether with respect to the human right to water (Obani and Gupta 2014) or with respect to ground water (Conti and Gupta 2014), one finds that there are multiple rules operating on the same jurisdiction or multiple rules operating for the same groups of people (for example, indigenous peoples; Gupta, Hildering, and Misiedjan 2014). This becomes very obvious when one uses a legal pluralism lens to examine the nature of water policy. Such a lens reveals that there are layers of inconsistencies in water policies as newer policies have been superimposed on older policies without necessarily compensating, consulting, or even informing those whose rights have been eroded over time (Bavinck and Gupta 2014). Such competing rules may be positive when there is contestation regarding how the rules should be appropriately designed; and they may be perceived as negative as when such contestation neither addresses the ecological nor the social dimension, but instead leads to counterproductive policies.

Toward an Action Framework

Achieving the Sustainable Development Goals will require a broadly encompassing action framework design that embodies mechanisms to address the main drivers and barriers to change. Implementing these principles of mainstreaming Sustainable Development Goals across scales and actors, dealing with the diversity of drivers at the different levels, and ensuring horizontal and vertical coherence is not an easy task. Below we elaborate on three keys sets of interdependent action mechanisms: developing human capacities, building institutional capacities, and designing interventions.

Developing Human Capacities

At the most fundamental level, achievement of Sustainable Development Goals will depend on taking action for building capacities at individual and organizational levels. Building capacities has many facets, including enhancing the understanding of how paradigms shape production, distribution, and consumption, as well as strengthening the development and transmission of knowledge, know-how, and experience.

On the first point, in the case of water, at the global level it is important to understand which water paradigms (for example, water as a human right; water as an economic good; water as a heritage) are compatible with problem solving and under what conditions. It is also important to have a comprehensive understanding of the water system; the role of green, blue, gray, black, rainbow water, and virtual water; and demand and supply that is consistent with ecosystem service maintenance and the sharing of best practices. At regional and national levels, similar exercises need to be carried out, but in addition to detailed mapping of resources, rights and use patterns are useful. At the local level, monitoring and reporting capacity are needed. All these call for greater human understanding and capabilities to understand the importance of rule-making powers at different levels.

Capacity development needs are a function of problem understandings. If we accept that a key challenge is the global-level paradigm of continued economic growth under a market-based system, then this is an area in which greater capacities are needed in order to demonstrate the conditions under which this paradigm is consistent with sustainable water use. If we accept that a key challenge affecting water is global climate change, then it is inevitable that there is a closer examination of the conditions under which production, consumption, and distribution solutions can be made more coherent with addressing the climate change problem.

Capacity development also has very practical aspects. At a practical level, there is a need for greater understanding of which plants use less water or can survive on salt water, and therefore to invest more in those plants. It might call for a greater understanding of the role of green water in national contexts and how to maintain or even enhance this role. It calls for better capacities to understand of role of infrastructure in managing and changing water flows and the impacts of that on ecosystem services. At the local and household levels, it calls for greater consciousness in purchase patterns, but also how the infrastructure for water is managed within the household.

Building Institutional Frameworks

Building on our human capacities, societies must ensure that institutional frameworks, entailing both principles and rule systems, as well as organizational frameworks, are set in place so that people and communities are empowered to draw upon their capacities in the pursuit of their development.

In terms of water, which is a typical multi-level issue (Gupta, Pahl-Wostl, and Zondervan 2013), such an overarching global framework would combine the development and environmental principles of Rio (Rio Declaration 1992) with human rights principles (UNGA 2010); development aid principles (Paris Declaration 2005); and water law principles (UN Watercourses Convention 1997; ILA 2004, Berlin Rules); and these have to be appropriately adopted at the multiple levels of governance. There is one global water law (UN Watercourses Convention 1997) and the UN Economic Commission for Europe (UNECE) Water Convention of 1992 is now also open to international participation; there are supporting treaties at multiple levels of governance; and there are dozens of river-basin organizations and community-based organizations—a wealth of bodies to build upon. It is also important to link water epistemic communities to water treaties (Gupta 2014b). The global water law—the UN Watercourses Convention (1997)—entered into force in 2014. This convention focuses on the need to equitably share water resources between riparians. In addition, the UNECE Watercourses Treaty and Protocol (1992, 1999) focuses more on quality aspects. We also have a draft treaty on groundwater law (International Law Commission 2008).

Thus, the global arena has the potential legal instruments, which need some tweaking and combining if they are to play a significant role in water governance and in achieving the Sustainable Development Goal on water. Groundwater law is one area that is rich in legal pluralism, and there is confusion regarding its implementation. At the same time, the water arena is also governed by rules of the World Trade Organization and the 3,000 or more bilateral and some multilateral investment treaties, which see water as an economic good. The private sector also has a role to play in water service provision. The UN General Assembly and the UN Human Rights Council have also adopted the human right to water and sanitation, and so we see a network of global legal institutions being developed to manage water (Dellapenna et al. 2013). UN-Water, a coordination body at the global level, is trying to coordinate the efforts of UN agencies and partners working on water issues.

At the regional level, the 300 transboundary river basins and aquifers are governed through a network of treaties, some of which date back several hundred years. Many of these treaties have set up an organizing framework of river basin and catchment organizations as a mechanism for managing their transboundary waters (Jaspers and Gupta 2014; Huitema and Meijerink 2014); many of these countries already heavily engage stakeholders in such organizations. The organizational framework also includes the establishment of local water rights as well as reporting, monitoring, and enforcement measures. At the national level, while about half of the Organisation for Economic Co-operation and Development (OECD) nations have water ministries controlling water policies, the other half have no dedicated water ministry, but see water as a crosscutting issue—one that cuts across the work terrain of other ministries, including those on industry, agriculture, forests and nature services, and the economy, if not the ministry of foreign affairs. Most developing countries have gone through several phases in their water policies through history, with an initial phase of customary law influenced by the rise of civilizations and religions. Subsequently, new water law has developed through conquest and colonization; in more recent years, water laws are based on paradigms such as communism (state-owned water), neo-liberalism (private-sector participation in water provision), or science, which calls for better understanding of the hydrological system, the relationship between climate change and water, the role of green water, the ecosystem services of water, and so on.

In the water field, we can conclude there is a wide range of existing institutions and instruments available that could be put in service of the Sustainable Development Goal on water. But effective action will require that the specific water targets are mainstreamed across different institutions and instruments, and that rule makers make a significant effort to build coherence across institutional frameworks at different levels.

Designing Interventions

The institutional framework provides the source of agency, power, and legitimacy to make interventions, which can be handled by a diversity of actors in society, including the state. For example, the recognition of water as a human right (UNGA 2010), when translated into national policy, can go a long way toward empowering citizens to assert their rights. Interventions can be regulatory (for example, strategic environmental assessments, spatial planning, standards); economic (for example, taxes, subsidies, insurances, micro-credit schemes, tariffs, soft loans, payment for ecosystem services); suasive (for example, public awareness, education, science, monitoring); management oriented (for example, corporate social responsibility, hybrid management, community management, certification schemes); or technical in nature (for example, drip irrigation, desalination plants, drought-resistant crops). Appropriate interventions by public and private actors should in turn enable the investments needed in different development sectors, and the financing arrangements for these investments.

Achieving Goal 6 on water and sanitation requires investment, but possibly not as much as many think. As well as investment, it requires a multi-level and multi-actor commitment to coherently achieve the goals of water management, maintaining the ecosystem services of water, equitably sharing water, and protecting water resource quality. Such coherence would also imply that instead of water policy working at cross purposes, resources are focused on achieving other key Sustainable Development Goals. For instance, since practically all social sectors use water, there is good reason to focus on public sources of finance. In many places, this in turn requires better domestic capacity for tax collection (Target 17.1; OECD 2014). It may be important to encourage private-sector participation, but with an important caveat. Private-sector participation in water management and water service delivery is often accompanied by confidential contracts. Where such contracts are made between international private-sector players and national authorities, these contracts are governed mostly by private international law or the rules of international investment law. Increasingly, such secretive contracts make transparent water governance more and more difficult, creating a potential inconsistency with the "good" governance targets such as Goal 16.6—"Develop effective, accountable and transparent institutions at all levels." Where such secretive contracts are contested, this can lead to secretive arbitration in international arbitration courts and takes the public good of water out of the public eye and into the realm of secrecy and confidentiality (Klijn, Gupta, and Nijboer 2009). While in the past, the key actors in dam infrastructure investment were large development banks that were sensitive to public opinion, the newer dams are being built by national banks and investment companies, leading to secret agreements regarding who will receive the water flows and at what costs (Merme, Gupta, and Ahlers 2014). Thus, to the extent that private-sector participation is encouraged in water governance, it is of vital importance that such contracts are transparent and open to public scrutiny and change as and when new knowledge becomes available.

Toward a Multi-level Action Framework for Water

It is not surprising that decision-makers, development planners, and practitioners become overwhelmed by the massive requirements for action if we are to deliver on the Sustainable Development Goals. Add to this the need to mainstream and monitor these goals in all relevant policies at multiple levels of governance and to ensure accountability, and the task appears insurmountable. However, organizing these requirements into a systematic action framework for each goal will help structure the work and allocate adequate resources to it. Table 12.1 presents in more detail the outline of an action framework for the goal on water, including different sets of action mechanisms at different scales.

This action framework makes it possible to return to issues of coherence, both vertically across levels and horizontally across mechanisms, in order to achieve the mainstreamed targets. Our concern with coherence across levels here is to ensure that, for example, capacity-building efforts or institutional arrangements are working toward the same goals across levels; and coherence across mechanisms aims to ensure that action mechanisms—whether organizational frameworks, financing arrangements, or policy instruments—are synergetic at each level. For example, capacity-building activities need to be calibrated against other mechanisms—such as building capacity specifically toward improved planning systems, public policy assessment, and leveraging investment finance.

Of particular interest for the UN system will be the monitoring and reporting frameworks that need to function across all scales—from local to national systems that speak with the global reporting framework.

Conclusion

This chapter has presented the key elements of a multilevel action framework for the Sustainable Development Goals, and illustrated this with the example of water. Our argument is based on the simple premise that detailed and comprehensive consideration of action mechanisms must be a central part of the Sustainable Development Goals throughout their implementation, underpinned by a shared understanding of key principles. We propose three such principles: first, mainstream Sustainable Development Goals as a multi-actor and multilevel governance question involving both state and nonstate actors. The state is accountable, so where nonstate actors cannot be mobilized, the state may have to do more; where nonstate actors can be mobilized, the state can do less. Second, make sure that the actions put in place address, at the appropriate scale, the direct and indirect drivers behind the challenges of the Sustainable Development Goals. Third, ensure horizontal and vertical coherence of actions.

These principles underpin an action framework based on three mechanisms: developing capacities, building institutional frameworks, and designing interventions. For example, coherence can be pursued through joint action agendas for different Sustainable Development Goals, such as policies on efficient use of resources in agriculture—working for water, food, and energy goals. This requires getting away from sector silos

Table 12.1

Examples of Action Mechanisms for the Sustainable Development Goal on Water across Levels

		Developing capacities	Building institutional frameworks	Designing interventions
Vertical integration	**Global**	- Discuss competing paradigms (for example, water as human right or economic good) and policy implications - Science reviews by, for example, the World Water Assessment Programme (2009) - Share best practices	- Combine global environmental and developmental principles with human rights and development aid principles - Link with UN and UNECE water treaties	- Regulatory: for example, human right to water, water standards - Economic: for example, water tax or subsidy; agricultural insurance - Suasive: water labeling for products - Investment: Global Green Fund, Global Environment Facility, development banks, corporate social responsibility for water companies
	Regional	Develop resource assessments and map water resources	Update transboundary water agreements and organizations with Sustainable Development Goals	- Regulatory: human rights, use standards - Economic: water taxes, subsidies, and insurance - Suasive: create awareness of the global water crises; water labeling for products using water; Regional Development Banks to take water Sustainable Development Goals into account
	National	Map resources and resource use; capacity building for managing and financing water	Revisiting water policy structure in country (one or many ministries; federal or provincial subject)	- Regulatory (including strategic environmental assessments, spatial planning), economic, suasive, management, technological - Investment (for example, national investment funds, disaster management funds, public-private partnerships)
	Local	Building monitoring and reporting capacity; building local capacity for integrated and development planning	Using or creating community-based organizations for water basin management; community-based organizations need to internalize the goal	- Integrating the Sustainable Development Goals with local and customary water rights; - Local investment funds; micro credit
		Horizontal integration		

and thinking about crosscutting actions—including interventions such as resource-efficiency schemes, innovation schemes, or gender equality schemes. Resolving conflicts among Sustainable Development Goals can to some extent be undertaken through institutional procedures such as strategic environmental assessments or integrated assessments. Actions deal with the multiple drivers at the different levels that are causing the problem. This might require that actions have to be driver-specific and level-specific; in other words, coherence in actions does not imply the actions have to be identical, but simply relevant to addressing the appropriate drivers at different levels of governance.

A central argument for the action framework is that one cannot start with interventions and investments unless human capacities and institutional frameworks are in place; in the most challenging country contexts, one has to spend most effort on capacity building and on building institutions and governance systems. For example, the one-sided promotion of market-based instruments as the key fix for various resource challenges, which has so often been the mainstream prescription from international financial institutions in the last 20 years, has proven ineffective in contexts where institutions that regulate and ensure that markets function are not established and where people cannot afford to pay for access to basic services. In many cases, they lead to increasing socioeconomic inequality.

On the one hand, we need action worldwide to deal with specific local problems. On the other hand, we need global collaboration to ensure that goals add up in terms of, for example, environmental and resource requirements at the global level. In other words, global goals and principles need to be followed by multilevel action to deliver on these goals and principles from global to local levels.

References

Alston, Philip. 2005. Ships Passing in the Night: The Current State of the Human Rights and Development Debate Seen through the Lens of the Millennium Development Goals. *Human Rights Quarterly* 27 (3): 755–829.

Bavinck, Maarten, and Joyeeta Gupta, eds. 2014. SI: Sustainability science. Special issue on legal pluralism, governance and aquatic systems. *Current Opinion in Environmental Sustainability* 11, 1–94.

Bello, Walden. 2013. Post 2015 Development Assessment: Proposed Goals and Indicators. *Development* 56 (1): 93–102.

Benson, Charlotte, and John Twigg. 2007. *Tools for Mainstreaming Disaster Risk Reduction: Guidance Notes for Development Organizations*. Prevention Consortium.

Bernstein, Steven, Joyeeta Gupta, Steinar Andresen, Peter M. Haas, Norichika Kanie, Marcel Kok, Marc A. Levy, et al. 2014. Coherent Governance, the UN and the SDGs. Policy brief 4. United Nations University Institute for the Advanced Study of Sustainability.

Chopra, Kanchan, Rik Leemans, Pushpam Kumar, and Henk Simons, eds. 2005. *Ecosystem Services and Human Well-being: Policy Responses. Millennium Ecosystem Assessment.* Vol. 3., 489–523. Washington, DC: Island Press.

Commission on Global Governance. 1995. *Our Global Neighbourhood: The Report of the Commission on Global Governance*. Oxford University Press.

Conti, Kirstin, and Joyeeta Gupta. 2014. Protected by Pluralism? Grappling with Multiple Legal Frameworks in Groundwater Governance. *Current Opinion in Environmental Sustainability* 14: 39–47.

Council of Europe. 1998. *Gender Mainstreaming: Conceptual Framework, Methodology, and Conceptualisation of Existing Practices*. Strasbourg: Council of Europe.

Dellapenna, Joseph W., Joyeeta Gupta, Wenjing Li, and Falk Schmidt. 2013. Thinking about the Future of Global Water Governance. *Ecology and Society* 18 (3): 28. Available at: http://www.ecologyandsociety.org/vol18/iss3/art28/.

Doornbos, Martin. 2001. "Good Governance": The Rise and Decline of a Policy Metaphor? *Journal of Development Studies* 37 (6): 93–108.

Dorsey, Ellen, Mayra Gómez, Bert Thiele, and Paul Nelson. 2010. Falling Short of Our Goals: Transforming the Millennium Development Goals into Millennium Development Rights. *Netherlands Quarterly of Human Rights* 28 (4): 516–522.

Dublin Declaration. 1992. The Dublin Statement on Water and Sustainable Development. Report of the International Conference on Water and the Environment.

ECOSOC, United Nations Economic and Social Council. 2006. Mainstreaming a Gender Perspective into All Policies and Programmes in the United Nations System. Resolution 2006/36. Available at: http://www.unhcr.org/refworld/docid/46c455acf.html.

European Commission. 2000. Directive 2000/60/EC of the European parliament and of the council of 23 October 2000 establishing a framework for community action in the field of water policy. *Official Journal of the European Communities* L327: 1–72.

Gupta, Joyeeta. 2008. Global Change: Analyzing Scale and Scaling in Environmental Governance. In *Institutions and Environmental Change: Principal Findings, Applications, and Research Frontiers*, ed. Oran R. Young, Leslie A. King, and Heike Schroeder, 225–258. Cambridge, MA: MIT Press.

Gupta, Joyeeta. 2010. Mainstreaming Climate Change: A Theoretical Exploration. In *Mainstreaming Climate Change in Development Cooperation: Theory, Practice and Implications for the European Union*, ed. Joyeeta Gupta and Nicolien van der Grijp, 67–96. Cambridge, UK: Cambridge University Press.

Gupta, Joyeeta. 2014a. Glocal Politics of Scale on Environmental Issues: Climate, Water and Forests. In *Scale-sensitive Governance of the Environment*, ed. Frans J. G. Padt, Paul F. M. Opdam, Nico B. P. Polman, and Catrien J. A. M. Termeer, 140–156. John Wiley and Sons.

Gupta, Joyeeta. 2014b. Science and governance: Climate change, forests, environment and water governance. In *The Role of Experts in International Decision-Making: Irrelevant, Advisors or Decision-Makers*, ed. Monika Ambrus, Karin Arts, Helena Raulus, and Ellen Hey, 148–170. Cambridge University Press.

Gupta, Joyeeta, Antoinette Hildering, and Daphina Misiedjan. 2014. Indigenous peoples and their right to water: A legal pluralism perspective. *Current Opinion in Environmental Sustainability* 11: 26–33.

Gupta, Joyeeta, and Claudia Pahl-Wostl. 2013. Global Water Governance in the Context of Global Governance in General and Multi-level Governance: Its Need, Form, and Challenges. *Ecology and Society* 18 (4): 53.

Gupta, Joyeeta, Claudia Pahl-Wostl, and Ruben Zondervan. 2013. "Glocal" Water Governance: A Multilevel Challenge in the Anthropocene. *Current Opinion in Environmental Sustainability* 5: 573–580.

Hooghe, Liesbeth, and Gary Marks. 2003. Unravelling the Central State, but How? Types of Multi-level Governance. *American Political Science Review* 97 (2): 233–243.

Huitema, Dave, and Sander Meijerink, eds. 2014. *The Politics of River Basin Organizations*. Cheltenham: Edward Elgar.

ILA, International Law Association. 2004. Berlin Rules. Report of the Seventy-First Conference, Berlin. London: International Law Association.

International Law Commission. 2008. Draft Articles on the Law of Transboundary Aquifers. Report of the International Law Commission on the Work of Its Sixtieth Session, UN GAOR, 62d Sess., Supp. No. 10, at 19, UN Doc. A/63/10.

Jaspers, Frank, and Joyeeta Gupta. 2014. Global Water Governance and River Basin Organizations. In *The Politics of River Basin Organizations*, ed. Dave Huitema and Sander Meijerink, 38–66. Edward Elgar.

Kaul, Inge, Pedro Conceicao, Katell Le Goulven, and Ronald U. Mendoza. 2003. Why Do Global Public Goods Matter Today? In *Providing Global Public Goods*, ed. Inge Kaul, Pedro Conceicao, Katell Le Goulven, and Ronald U. Mendoza, 2–20. Oxford: Oxford University Press.

Klijn, Anne-Marie, Joyeeta Gupta, and Anita Nijboer. 2009. Privatising Environmental Resources: The Need for Supervision. *Review of European Community & International Environmental Law* 18 (2): 172–184.

Kyoto Protocol. 1997. Kyoto Protocol to the United Nations Framework Convention on Climate Change. Signed 10 December 1997, in Kyoto; entered into force 16 February 2005. Reprinted in (1998). *International Legal Materials* 37 (1): 22.

Liberatore, Angela. 1997. The Integration of Sustainable Development Objectives into EU Policy Making: Barriers and Prospects. In *The Politics of Sustainable Development*, ed. Susan Baker, Maria Kousis, Dick Richardson, and Stephen Young, 104–123. Routledge.

Marks, Gary, Liesbeth Hooghe, and Kermit Blank. 1996. European Integration from the 1980s State Centric vs. Multilevel Governance. *Journal of Common Market Studies* 34 (3): 341–378.

Merme, Vincent, Joyeeta Gupta, and Rhodante Ahlers. 2014. Private Equity, Public Affair: Hydropower Financing in the Mekong Basin. *Global Environmental Change* 24: 20–29.

Nilsson, Måns, and Åsa Persson. 2012. Can Earth System Interactions Be Governed? Governance Functions for Linking Climate Change Mitigation with Land Use, Freshwater and Biodiversity Protection. *Ecological Economics* 75: 61–71.

Nilsson, Måns, Tony Zamparutti, Jan Erik Petersen, Björn Nykvist, Peter Rudberg, and Jennifer McQuinn. 2012. Understanding Policy Coherence: Analytical Framework and Examples of Sector-Environment Policy Interactions in the EU. *Environmental Policy and Governance* 22: 395–423. doi: 10.1002/eet.1589.

Obani, Pedi, and Joyeeta Gupta. 2014. Legal Pluralism in the Area of Human Rights: Water and Sanitation. *Current Opinion in Environmental Sustainability* 11: 63–70.

Organisation for Economic Co-operation and Development (OECD). 2014. Tax Transparency 2014, Report on Progress. Global Forum on Transparency and Exchange of Information for Tax Purposes.

Paris Declaration. 2005. Paris Declaration on Aid Effectiveness: Ownership, Harmonisation, Alignment, Results and Mutual Accountability. High-level Forum on Aid Effectiveness, February 28–March 2, 2005, Paris. Available at: http://www.oecd.org.

PBL. 2011. A Global Public Goods Perspective on Environment and Poverty Reduction: Implications for Dutch Foreign Policy. The Hague.

Redondo, Elvira D. 2009. The Millennium Development Goals and the Human Rights Based Approach: Reflecting on Structural Chasms with the United Nations System. *International Journal of Human Rights* 13 (1): 29–43.

Rhodes, Roderick A.W. 1996. The New Governance: Governing without Government. *Political Studies* 44: 652–667. doi: 10.1111/j.1467-9248.1996.tb01747.x.

Rio Declaration. 1992. Rio Declaration and Agenda 21. Report on the UN Conference on Environment and Development, Rio de Janeiro, 3–14 June 1992, UN Doc. A/CONF.151/26/Rev.1 (Vols.1-III).

Robinson, Mary. 2010. The MDG-Human Rights Nexus and Beyond 2015. *IDS Bulletin* 41 (1): 80–82.

Rockström, Johan, Will Steffen, Kevin Noone, Åsa Persson, F. Stuart Chapin, Eric F. Lambin, Timothy M. Lenton, et al. 2009. A Safe Operating Space for Humanity. *Nature* 461: 472–475.

Rosenau, James N. 1992. Governance, Order, and Change in World Politics. In *Governance without Government: Order and Change in World Politics*, ed. James N. Rosenau and Ernst-Otto Czempiel, 1–29. Cambridge, UK: Cambridge University Press.

Sanwal, Mukul. 2012. Global Sustainable Development Goals: The Unresolved Questions for Rio+20. *Economic and Political Weekly* xlvii: 14–16.

Scharpf Fritz, W. 1997. *Games Real Actors Play. Actor-centered Institutionalism in Policy Research*. Westview Press.

Stiglitz, Joseph. 2000. Vers un nouveau paradigme du développement. *Economics and Politics* 5: 6–39.

UN, United Nations. 2012. The Future We Want. Outcome document of the UN Conference on Sustainable Development (Rio+20), Rio de Janeiro, Brazil. UN Doc. A /CONF.216/L.1.

UNDP, United Nations Development Programme. 1999. Global Public Goods: International Cooperation in the 21st Century. New York.

UNDP, United Nations Development Programme. 2014. Human Development Report 2014. Sustaining Human Progress: Reducing Vulnerabilities and Building Resilience.

UNECE, United Nations Economic Commission for Europe. 1992. Convention on the Protection and Use of Transboundary Watercourses and International Lakes. UN Reg. 33207, 1936 UNTS 269.

UNECE, United Nations Economic Commission for Europe. 1999. London Protocol on Water and Health on the Convention on the Protection and Use of Transboundary Watercourses and International Lakes (25 Parties).

UNEP, United Nations Environment Programme. 2013. Global Environmental Outlook 5.

UN Framework Convention on Climate Change. 1992. United Nations Framework Convention on Climate Change. *International Legal Materials* 31 (4): 849.

UNGA, United Nations General Assembly. 2010. Resolution on Human Right to Water and Sanitation. UN Doc. A/64/292. Available at: http://www.un.org/News/Press/docs/2010/ga10967.doc.htm.

UNGA, United Nations General Assembly. 2015. Transforming Our World: The 2030 Agenda for Sustainable Development. Draft resolution referred to the United Nations summit for the adoption of the post-2015 development agenda by the General Assembly at its sixty-ninth session. UN Doc. A/70/L.1.

UNIDO, United Nations Industrial Development Organization. 2008. Public Goods for Economic Development. Vienna.

UN Watercourses Convention. 1997. Convention on the Law of the Non-Navigational Uses of International Watercourses. UN Doc. A/51/869, adopted in Res. A/RES/51/229.

Vörösmarty, Charles J., Peter McIntyre, Mark O. Gessner, David Dudgeon, Alexander Prusevich, Pamela A. Green, Stanley Glidden, et al. 2010. Global Threats to Human Water Security and River Biodiversity. *Nature* 467: 555–561.

Walby, Sylvia. 2005. Gender Mainstreaming: Productive Tensions in Theory and Practice. *Social Politics* 12 (3): 321–343.

Weitz, Nina, Måns Nilsson, and Marion Davis. 2014. A Nexus Approach to the Post-2015 Agenda: Formulating Integrated Water, Energy, and Food SDGs. *SAIS Review (Paul H. Nitze School of Advanced International Studies)* 34 (2): 37–50. doi: 10.1353/sais.2014.0022.

Went, Robert C.P.M. 2010. *Internationale Publieke goederen: Karakteristieken en typologie*. The Hague.

World Water Assessment Programme. 2009. *The United Nations World Water Development Report 3: Water in a Changing World*. Paris: UNESCO Publishing.

Young, Oran R., Arild Underdal, Norichika Kanie, Steinar Andresen, Steven Bernstein, Frank Biermann, Joyeeta Gupta, Peter M. Haas, Masahiko Iguchi, Marcel Kok, Marc Levy, Måns Nilsson, László Pintér, and Casey Stevens. 2014. Earth System Challenges and a Multi-layered Approach for the Sustainable Development Goals. Post 2015/UNU-IAS Policy Brief #1.

13 Conclusion: Key Challenges for Global Governance through Goals

Frank Biermann and Norichika Kanie

As numerous commentators have suggested, 17 Sustainable Development Goals can be seen as an unprecedented step to advance and strengthen a novel type of governance that will guide and "orchestrate" public policies and private efforts over the next 15 years. While there have been predecessors—notably the Millennium Development Goals—the new Sustainable Development Goals are different and unique in their broad coverage and specific characteristics. Yet, as critics also rightfully pointed out, it is far too early to fully assess the eventual effectiveness of this new approach of setting global universal goals on sustainable development. At the time of this writing, the Sustainable Development Goals are barely one year old. In chapters 1–12 of this volume, our group has thus offered nothing more than a first analysis and assessment of the evolution, rationale, and future prospects of the Sustainable Development Goals as examples of a novel type of "governance through goals."

In this concluding chapter, we summarize some key findings of this volume, with a view to general implications for governance through goals as a novel mechanism of world politics. In addition, we discuss the challenges for, and opportunities of, the Sustainable Development Goals by identifying several conditions that might determine their successful implementation. We also suggest some possible avenues for further research.

Governing through Goals

As emphasized throughout this volume, the approach of "global governance through goals"—and the Sustainable Development Goals as a prime example—is marked by a number of key characteristics. None of those is specific to this type of governance. Yet all these characteristics together, in our view, amount to a unique and novel way of steering and distinct type of institutional arrangement in global governance.

First, it is important to note that governance through goals is detached from the international legal system. The Sustainable Development Goals are not legally binding, and the instrument that established the goals at the global level—a resolution by the UN General Assembly—is in no way intended to grant legal force to the goals. Accordingly, there is no further ratification process, and governments are under no legal obligation to formally transfer the goals into their national legal systems. This distinguishes these types of goals from other goals or targets on sustainable development issues that have been clearly enshrined in legally binding treaties, for example under the regimes on the protection of the stratospheric ozone layer or marine environment. However, this also does not prevent some Sustainable Development Goals from possibly becoming part of legal regimes that are agreed upon elsewhere, including subglobal legal systems. For example, Goal 13 on climate change is essentially a referral to the legally binding climate convention and its 2015 Paris Agreement.

Instead of legal systems, measurability is at the core of governing through goals. By measuring progress using indicators, countries and actors can be compared. Leaving actors free to attain goals and targets, individually or by utilizing networks, while measuring and comparing their progress is a unique characteristic of governance through goals.

Second, governance through goals, as exemplified by the Sustainable Development Goals, functions through weak institutional arrangements at the intergovernmental level. These arrangements are different, for example, from the complex institutional mechanisms that have been created for the more specific governance domains of climate stability, ozone layer protection, or biodiversity preservation. The institutional oversight over implementation of the Sustainable Development Goals at the global level has been left rather vague, and will now be fulfilled by a High-level Political Forum on Sustainable Development that in itself is new and still has to prove its effectiveness (see in more detail Bernstein, this volume, chapter 9; Biermann 2014). Sustainable Development Goal 17 addresses broadly the need to build up a global partnership for implementing the Sustainable Development Goals, somewhat comparable to the Millennium Development Goal 8. However, there are also fundamental differences between both goal-setting processes. The Millennium Development Goal 8, for instance, was less focused on voluntary commitments than the current discourse around Sustainable Development Goal 17 seems to suggest. In addition, global partnerships for specific goals—and possibly even more so for specific targets—will most likely emerge, and some are already in existence, such as on health, water, oceans, and sustainable consumption and production. All these issue-specific partnerships can provide some global oversight and steering mechanisms that can support implementation of the goals over the next 15 years.

Weak global institutional arrangements, however, do not necessarily imply a low likelihood of successful implementation of the goals. It is especially the bottom-up, nonconfrontational, country-driven, and stakeholder-oriented aspects of governance through goals that its supporters cite as a key success factor for these goals, given that they had to be agreed—and will have to be implemented—in a highly diverse and divided state system where global multilateral agreement on detailed legally binding standards has proven to be difficult in recent years.

Third, interestingly, the Sustainable Development Goals are not the result of a broader strategic planning process at the global level. As we laid out in chapter 1 in more detail, the first proposal essentially dates back to only 2011. When the 2012 UN Conference on Sustainable Development followed this proposition a year later and decided to elaborate a set of Sustainable Development Goals by 2015, few would have expected this mechanism to become the central element of the post-2015 development agenda. Of course, there had been some previous discussion on how to address the remaining tasks and challenges of the Millennium Development Goals that were to expire in 2015. And in May 2013, the UN Secretary-General's High-level Panel of Eminent Persons on the Post-2015 Development Agenda submitted an indicative list of 12 goals and associated targets that helped to set the pace for the following discussion on the post-2015 development agenda. And yet this was seen in the beginning rather as a process for "post Millennium Development Goals," with limited goals and targets; few have foreseen the central role that the Sustainable Development Goals have eventually taken within the UN system. In a sense, the Sustainable Development Goals have become the central element of the 2030 agenda almost by coincidence, owing to the evolution of the global negotiations around 2014 when the Open Working Group on the Sustainable Development Goals became the only major intergovernmental forum to discuss the post-2015 development agenda.

Fourth, the new approach of governance through goals works through global inclusion and comprehensiveness of the global goal-setting process. Unlike the earlier Millennium Development Goals, the new Sustainable Development Goals address both industrialized and developing countries. Conceptually, this approach also turns countries in North America, Europe, East Asia, and Oceania into "developing countries" that have to bring forward plans to transform their societies toward more sustainable development paths. Again, unlike the earlier Millennium Development Goals, the new Sustainable Development Goals cover the entire breadth of the sustainability domain to more or less equal extent, thus combining a focus on economic development and poverty eradication (the core of the Millennium Development Goals) with a strong concern for social justice, better governance, and environmental protection and resource efficiency. This is an important development in global agenda setting. Even though conferences such as the 1992 UN Conference on Environment and Development, the 2002 World Summit on Sustainable Development, and the 2012 UN Conference on Sustainable Development were attended by numerous heads of state and government, sustainable development as such had not been on the main agenda in UN affairs before—as evidenced, for example, by the relatively low practical standing of the Commission on Sustainable Development in the UN system. Yet with the Sustainable Development Goals, this might now have changed, and sustainable development might now have received a status that puts it on an equal footing with the more traditional international economic agenda. With the new Sustainable Development Goals, after 40 years, the integrative concept of "sustainable development" has now finally become the core agenda for the United Nations with its new strategy of global governance through goals.

Fifth, inclusion and comprehensiveness also extends to the origin of the new Sustainable Development Goals in the period 2012–2015. While the earlier Millennium Development Goals were essentially elaborated within the UN Secretariat—even though in the context of the Millennium Declaration and with various inputs by governments (Loewe 2012; Manning 2010; McArthur 2014)—the new Sustainable Development Goals were agreed upon in a public process that involved input from at least 70 governments as well as numerous representatives of civil society. It is important to note that this process of formulating the goals was spearheaded by middle-income countries from the South—especially Colombia and Guatemala—and could draw on broad participation by civil society, including numerous online fora. In short, this is not a list of goals made up by bureaucrats selected by the UN Secretary-General, as the Millennium Development Goals have essentially been.

Sixth, as a consequence of the preceding points, governance through goals, as exemplified by the Sustainable Development Goals, grants much leeway to national choices and preferences. Even though no fewer than 169 targets have been agreed upon at the global level to guide implementation of the 17 Sustainable Development Goals, many of these targets are qualitative and leave maximum freedom for governments to determine their own ambition in implementing the goals. Even when quantitative and clearly defined targets have been chosen—including, for example, the goal of completely freeing the world from hunger by 2030—governments can still rely on the nonbinding nature of the goals, or take recourse to lack of support by the international community in fulfilling the goals and targets. So, in effect, governments retain maximum freedom in interpreting and implementing the goals if they so choose. This inbuilt leeway helps to explain why, for example, countries that constitutionally enshrine the discrimination of women and girls have still supported a global goal on achieving gender equality (Goal 5), or why countries with strong neoliberal economic beliefs could agree on a goal to reduce national and international inequality (Goal 10). In the end, this relatively flexible approach allowed all governments to agree on the complete package of 17 Sustainable Development Goals and 169 targets, turning them into a universal vision and guideline for public policy over the next 15 years.

Yet the character of the new Sustainable Development Goals as nonbinding global aspirations with weak institutional oversight arrangements and high levels of national discretion does not imply that our book concludes with an outright negative, pessimistic assessment. Instead, we do see potential for a global governance strategy through goals, as represented by the Sustainable Development Goals, to advance public policy and private efforts toward an ambitious sustainability agenda—admittedly also in light of the lack of alternatives given the current state of global governance and intergovernmental policies that are so far insufficiently responding to the challenges we face in the Anthropocene (Biermann 2014; Young et al., this volume, chapter 3). Much will depend on the future policy development around these goals over the next years, from the evolution of the global institutional arrangements to the ambition of the eventual national and subnational implementation process. Following this line of reasoning, in the following sections we lay out several conditions that could help, we believe, the new goals turn into a success story.

Challenges in Implementation

Further Strengthening the Goals through Indicators and Commitments

To start with, even though the 17 Sustainable Development Goals are supported by no fewer than 169 more concrete targets, many of these targets remain comparatively vague. Most are also purely qualitative, leaving much room for interpretation and hence weak implementation.

For this reason, it will now be important to concretize the Sustainable Development Goals as much as possible through appropriate indicators, combined with formalized commitments by governments at national level. As Oran Young (this volume, chapter 2) has pointed out in more detail, the success of governing through goals depends on the increasing formalization of commitments, the establishment of clear benchmarks, and the issuance of formal, measurable pledges by governments, all of which may, in the words of Young, "cause embarrassment or loss of face" in case of noncompliance. Just as the regime-building toward the 2015 Paris Agreement under the climate convention started with national pledges (so-called Intended Nationally Determined Contributions), a reverse approach of governance through goals could expect that national-level commitments now follow the broad global agreement of the Sustainable Development Goals.

In addition to national commitments, the struggle for a demanding governance system continues also at the level of indicators that must effectively support the broad ambition expressed in the Sustainable Development Goals. Ideally, this process will go beyond traditional means of national reports and reviews and include other types of review systems (Bernstein, this volume, chapter 9). One question will be whether the "ecumenical diversity and soft priorities" (Underdal and Kim, this volume, chapter 10) evidenced in the 17 Sustainable Development Goals will be sufficient to effectively guide behavior and set priorities—without an integrating vision and principle of what long-term sustainable development in the Anthropocene means. This problem led Underdal and Kim to forcefully argue for a basic norm (*Grundnorm*) of sustainability that would provide such much-needed prioritization and "orchestration" in implementing the Sustainable Development Goals.

Strengthening Global Governance Arrangements

In addition, a major challenge for implementation of the Sustainable Development Goals rests on the appropriate governance arrangements at the global level, including on how those are integrated into, or aligned with, existing institutions. As pointed out by Young (this volume, chapter 2), devising effective procedures to track progress is a key element of success for a global governance strategy that relies on goal-setting. Importantly, unlike international legal regimes, governance through goals starts with aspirations that are not necessarily coupled with existing or new governance arrangements (see also Underdal and Kim, this volume, chapter 10; Gupta and Nilsson, this volume, chapter 12).

One central new body here is the new High-level Political Forum on Sustainable Development. As discussed in detail in chapters 1 and 9 in this volume, governments agreed in 2012 on creating this forum, and in 2015 gave the forum "a central role in overseeing a network of follow-up and review processes at the global level" (UNGA 2015, par. 82) for the Sustainable Development Goals.

One element of monitoring progress on the Sustainable Development Goals will be a new global sustainable development report that should be, following the United Nations, forward-looking, policy-oriented, and reporting not only on areas of success but also on policy gaps and shortcomings. The report shall serve as a key element in the science-policy interface under the High-level Political Forum, as a key instrument in monitoring the road between 2015 and 2030 (Bernstein, this volume, chapter 9). And yet, in 2012, the details and function of the High-level Political Forum were not clearly laid out, and hence have been subject to further intergovernmental negotiations within the United Nations since then (Bernstein, chapter 9, this volume). Similarly, how the High-level Political Forum could function as an "orchestrator" in global sustainability governance remains an open question (Bernstein, chapter 9, this volume; Underdal and Kim, chapter 10, this volume).

In addition, key for the success of the Sustainable Development Goals is reliable and predictable resource mobilization (Bernstein, this volume, chapter 9; Voituriez et al., this volume, chapter 11). Sustainable Development Goal 17 is explicitly targeted at a "global partnership." This global partnership will need to include additional funding from public sources, as has been the case for the global partnership under the Millennium Development Goals. Yet public-private and private-private partnerships and other types of action networks will also be important, probably more so than they have been for the Millennium Development Goals. Since the 2012 UN Conference on Sustainable Development, 2,110 voluntary commitments have been made. Their success is difficult to predict. Much will depend, for instance, on institutionalized review mechanisms and clear and quantifiable benchmarks that measure performance under the global partnership to implement the Sustainable Development Goals (Bernstein, this volume, chapter 9; Voituriez et al., this volume, chapter 11). In addition, leadership of individual actors—including pioneering countries, as has been the case of Norway in the area of health governance—might be crucial in specific circumstances (Andresen and Iguchi, this volume, chapter 7).

Partnerships do not need all to be global in scope. The specific character of the Sustainable Development Goals leaves much space for flexibility in implementation; this also makes it likely that much of the implementation will occur through more limited partnerships among fewer countries that probably will also include nonstate actors. Such partnerships can also cover only a small number of actors in issue-specific "coalitions of the willing" that might focus on only specific targets. One potential example of this kind of partnership is the recent development of "Champions 12.3," a global coalition of actors that seeks to help implement target 12.3, to cut in half per capita global food waste at the retail and consumer level and reduce food losses along production and supply chains (including post-harvest losses) by 2030 (http://champions123.org; see also Andresen and Iguchi, this volume, chapter 7). A company-level initiative has been recently taken, for example, by Unilever. An example of national-level stakeholder initiative is the "OPEN 2030" project in Japan, which tries to support the Sustainable Development Goals through a multi-stakeholder partnership (http://open2030project.com). The 2030 Agenda for Sustainable Development also encourages opportunities for learning, including through reviews, sharing best practices, and creating online platforms at regional and global levels.

On the one hand, problem solving through such partnerships may better fit complex problems as they are typical in the Anthropocene. Complex teleconnections and the nonlinear nature of the problems (which are specific features of the Anthropocene; see Biermann 2014 and Young et al., this volume, chapter 3) might be better addressed in a manner that leaves room for maneuver and rapid adjustment in more flexible governance arrangements that might be focused on particular goals (Kanie et al. 2014). On the other hand, such novel partnerships around the Sustainable Development Goals must not repeat the mistakes and failures of the many multisectoral, public-private partnerships that were agreed upon as "type II outcomes" of the 2002 World Summit on Sustainable Development in Johannesburg (Biermann et al. 2007; Bäckstrand et al. 2012). In short, much will thus depend on the effectiveness of the emerging network of implementation mechanisms and partnerships that will emerge to help turn the ambitious Sustainable Development Goals into concrete progress by 2030.

Adapting Global Ambitions to National Circumstances and Priorities

The Sustainable Development Goals aspire for universal application and are thus global in nature. Yet they are also expected to be adapted to the national and local context by taking into account the national reality, capacity, and other national contexts, including level of development and existing national and local policies. This is a significant departure from the earlier Millennium Development Goals, which had been set at the global level and were hence often criticized as a "one-size-fits-all" approach (Andresen and Iguchi, this volume, chapter 7). Yet as Gupta and Nilsson highlight (this volume, chapter 12), the "translation" of global and mid-term (2015–2030) aspirations into national policies requires significant capacities at the national level, including functioning governance systems. Partially for this reason, governance in itself has now become a specific Sustainable Development Goal, with a number of targets that call upon governments to improve their own performance in measurable ways (Biermann et al., this volume, chapter 4).

Most countries have existing national development goals and plans that could be modified or upgraded to take account of the new global aspirations laid down in the Sustainable Development Goals. Such modification, however, will also need to include political debate and decisions over the vision, aspirational role, and position of a country, including the allocation of its domestic resources. For example, what should be the concrete target for countries as diverse as Norway, Nicaragua, or Nepal with regard to the global target to halve food waste per capita by 2030 (as set out in target 12.3 under Goal 12)? Should they go for a national waste reduction target per capita that equals the global average, or should they have deeper targets—or less demanding ones? The Sustainable Development Goals could thus help support the design of national policies within a framework of globally agreed ambitions and vision.

A key to linking the global aspirations as they are laid out in the 17 Sustainable Development Goals and their national adaption is the measurement of progress. This will require clear and broadly accepted indicators (Pintér et al., this volume, chapter 5). Once policies and measures by different countries that vary in wealth, context conditions, and priorities can be assessed by the same indicators, progress can be globally compared; international "naming and shaming" can then help motivate countries to nudge their programs forward.

The same holds in situations where a country adopts a similar kind of national target but in a different timeframe, or when a country relies on different criteria, or when a national target is much lower than the global aspiration now laid out in the Sustainable Development Goals. Goals 16 and 17 address the process through which translation between global and national goals and targets could be formulated. Yet again, these processes will also require national adaptation and adjustment.

Successful implementation thus requires effective translation between global aspirations and national contextual policies and/or aspirations. Potential pitfalls are the broad selectivity of the Sustainable Development Goals when addressed in national policy development. Some initial studies have been conducted in Sweden, the Netherlands, and Japan to contextualize the Sustainable Development Goals in national situations (on Sweden, see Weitz et al. 2015; on Japan, see POST2015 2016; on the Netherlands, Lucas et al. 2016). Some developing countries have taken steps ahead, especially those that have a national mechanism for the Millennium Development Goals in place that they can now adapt for the Sustainable Development Goals. Although the fundamental difference between the Millennium Development Goals and the Sustainable Development Goals must be recognized, some experiences in implementation, along with institutional arrangements in some cases, can be a powerful resource to draw on.

The same applies for the level of intergovernmental institutions and organizations, notably the United Nations. Now that clear goals and targets have been set at the UN level, all UN agencies and the supporting national development agencies can start moving forward toward streamlining their funding and policies in accordance with the Sustainable Development Goals.

However, there is no doubt that the success of the Sustainable Development Goals will not be possible with government action alone. As pointed out in the UN General Assembly declaration, the 2030 Agenda for Sustainable Development addresses all stakeholders, not only governments. As forcefully argued by Gupta and Nilsson (this volume, chapter 12), implementation of Sustainable Development Goals must be undertaken by all actors in society, in a process that is eventually "owned" by all. Thus, civil society organizations, the private sector, and even each individual citizen are called upon to work toward implementation of the goals. Ideally, this might even lead to a new type of social movement for the Sustainable Development Goals or for individual goals, as suggested by Young (this volume, chapter 2). In addition, the private sector might need to play a more active role than in the past. For the case of water governance, for instance, Yamada (this volume, chapter 8) argues that the success of goal-based governance in this area will largely depend on the level of corporate engagement with governments, and that the UN Global Compact will play an important role in this regard.

Ensuring Effective Policy Integration in Implementation

Effective implementation of the Sustainable Development Goals requires in many cases systems for issue-oriented problem solving that go beyond existing frameworks and institutions. A close eye on interlinkages is important here, with a view toward an integrated approach for implementation of the Sustainable Development Goals, as emphasized by several studies (Griggs et al. 2014; Sachs 2015; Stafford-Smith et al. 2016) and the Global Sustainable Development Report (UN DESA 2015). All argue for an integrated implementation of the Sustainable Development Goals to avoid negative trade-offs and to create synergies (see also Haas and Stevens, this volume, chapter 6; Gupta and Nilsson, this volume, chapter 12).

This is a prerequisite for the success of a global governance strategy for sustainability as such. There is no doubt that all concerns addressed under the eight Millennium Development Goals have been of utmost importance for development, with the overall great success stories in areas such as poverty eradication and prevention of hunger and malnutrition (even though attribution of these successes to the existence of these goals remains debatable). Therefore, it is vital that these primary concerns have primacy of place again in the Sustainable Development Goals, with the central ambition of freeing the world of hunger within the next 15 years. Yet equally important is the preservation of fundamental life-supporting functions of planet Earth. For example, all success in poverty eradication under the previous development programs might be negated if the ambitious goals under the climate convention and its 2015 Paris Agreement are not met. Only if we manage the earth's life-support systems—some of which have now crossed or are close to crossing so-called "planetary boundaries"—are further sustainable improvements in human development possible. Social sustainability and social justice are crucial variables as well. In the case of food security, for example, theoretically there is enough food to feed all people on earth: The key problem is not the absolute amount of food, but its distribution and management.

For these reasons, an integrated approach for the three dimensions of sustainable development is indispensable. Such integration is required at all levels of global governance, from global to regional, national and local levels, and cutting across sectoral borders. For most of the countries this will require, for example, a reorganization of their national administrations and government systems. Integration in research also requires more interdisciplinarity, the breaking down of the silos of disciplinary knowledge, and the development of novel types of transdisciplinarity that combine specialist and stakeholder expertise, along with better architectures for an effective science-policy interface.

Improving the Adaptability of Governance Mechanisms

A final condition for the successful implementation of the Sustainable Development Goals is the adaptability of the related governance arrangements to deal with changes that are likely to take place over the next 15 years. Governance through goals in this regard may have to be more flexible to adjust to new situations. Yet how such flexibility will be maintained after now-emerging further institutionalization remains an open question.

For example, we have witnessed numerous changes since 2001 when the Millennium Development Goals were established. The economies of countries such as China and India have grown rapidly, which has lifted millions after millions of people out of abject poverty—yet also further increased local environmental pollution combined with growing global emissions of carbon dioxide and other pollutants. The unequal speed of development eventually resulted in the further diversification of interests among developing countries, which limited the coherence of the Group of 77, their central coalition in multilateral negotiations. Progress in science and the development of better models and scenarios of the earth system showed the need to change human behavior in order to avoid catastrophic events; and gradually the knowledge was shared. More nonstate actors participate in decision making at various levels of governance than before. Thanks to the rapid development of information and communication technologies and social media, citizens are now linked and networked with each other at a speed much faster than ever before. Many of these recent developments, however, were not accurately predicted in 2000. One simple example is the approach in the Millennium Development Goals to use the percentage of people with landline telephone connections as an indicator for communication and information improvements—which became quickly obsolete by the spread of mobile communication in barely a decade. The next 15 years will undoubtedly experience similarly major changes. Governance arrangements and core institutions for the attainment of the Sustainable Development Goals must thus be dynamic and flexible enough to respond to unpredictable changes over the next 15 years and beyond. Incorporating foresight processes is hence needed.

Outlook for Future Research Questions

Taken together, the information requirements posed by Sustainable Development Goals remain huge. These goals are a novel governance mechanism that is posing numerous new sets of questions for academic research and policy analysis.

To start with, the success of the Sustainable Development Goals will stand or fall with our *ability to measure*. To "measure what you treasure," however, is not trivial. A key task lies here with the international and national communities of statisticians, but also many other research communities are required (Pintér et al., this volume, chapter 5). For example, how can you measure progress in better governance, more transparent policies, less corrupt administrations, or better rule of law—all elements of Sustainable Development Goal 16—without further efforts in improving the methods underlying the appropriate indicators, along with increasing intergovernmental agreement on what indicators are most meaningful in assessing process (Biermann et al., this volume, chapter 4)?

Second, the new approach of governance through goals poses important new research questions regarding the *embedding and integration of goals at the global level* into existing governance arrangements; the *effects the goals at the global level may have* on other governance systems; and the question of *to what extent further governance reforms are needed* to cope with the resulting challenges. We have touched on these questions in a number of chapters (for instance, chapters 3, 8, 9, and 12), but see great further research needs in this field. "Orchestration" in global governance might be one overarching concept to understand the function of the Sustainable Development Goals—even though some might argue that a better description for governance through goals might even be conductorless jazz, given the bottom-up nature and emerging properties within a common vision!

Important will also be the *academic support for the integration of the economic, social, and environmental dimensions* of the Sustainable Development Goals. While the Millennium Development Goals were essentially related to a traditional development agenda, and other goal-setting processes have prioritized, for example, environmental concerns, the Sustainable Development Goals now attempt to integrate all three traditional dimensions of sustainable development. Essentially, all 17 Sustainable Development Goals touch upon all three dimensions at the same time, though to different degrees. This is even truer when acting to attain a goal. Integrating these dimensions with their different agendas and rationales in the implementation of these goals is a key challenge for decision-makers and other stakeholders at all levels of governance; yet it is also an important issue that the research communities, in interdisciplinary and transdisciplinary research projects, need to address. The emerging focus of the research community on the food-water-energy nexus, for example, reflects the importance of an integrated approach for sustainability, as well as a stronger focus on the social dimension.

In the end, there is no doubt that the Sustainable Development Goals pose one of the most ambitious, but at the same time also most daunting political challenges of our time—for both global and local governance. As the UN Secretary-General aptly summarized after conclusion of the 2030 Agenda for Sustainable Development and the 2015 Paris Agreement on climate: "We are the first generation that can end poverty, and the last one that can take steps to avoid the worst impacts of climate change. With the adoption of a new development agenda, sustainable development goals and climate change agreement, we can set the world on course for a better future." Novel types of governance through goals, we believe, will certainly be a vital part of this ambitious agenda.

References

Bäckstrand, Karin, Sabine Campe, Sander Chan, Ayşem Mert, and Marco Schäferhoff. 2012. Transnational Public-Private Partnerships. In *Global Environmental Governance Reconsidered*, ed. Frank Biermann and Philipp Pattberg, 123–147. Cambridge, MA: MIT Press.

Biermann, Frank. 2014. *Earth System Governance: World Politics in the Anthropocene*. Cambridge, MA: MIT Press.

Biermann, Frank, Man-san Chan, Ayşem Mert, and Philipp Pattberg. 2007. Multi-stakeholder Partnerships for Sustainable Development: Does the Promise Hold? In *Partnerships, Governance and Sustainable Development. Reflections on Theory and Practice*, ed. Pieter Glasbergen, Frank Biermann and Arthur P. J. Mol, 239–260. Cheltenham: Edward Elgar.

Griggs, David, Mark Stafford Smith, Johan Rockström, Marcus C. Öhman, Owen Gaffney, Gisbert Glaser, Norichika Kanie, Ian Noble, Will Steffen, and Priya Shyamsundar. 2014. An Integrated Framework for Sustainable Development Goals. *Ecology and Society* 19 (4): 49.

High-level Panel of Eminent Persons on the Post-2015 Development Agenda. 2013. *A New Global Partnership: Eradicate Poverty and Transform Economies through Sustainable Development*. New York: United Nations.

Kanie, Norichika, Steinar Andresen, and Peter M. Haas, eds. 2014. *Improving Global Environmental Governance: Best Practices for Architecture and Agency.* Oxon and New York, NY: Routledge.

Loewe, Markus. 2012. *Post 2015: How to Reconcile the Millennium Development Goals (MDGs) and the Sustainable Development Goals (SDGs)?* Bonn: German Development Institute.

Manning, Richard. 2010. The Impact and Design of the MDGs: Some Reflections. *IDS Bulletin* 41 (1): 7–14.

McArthur, John W. 2014. The Origins of the Millennium Development Goals. *SAIS Review (Paul H. Nitze School of Advanced International Studies)* 34 (2): 5–24.

Lucas, Paul, L. Kathrin Ludwig, Marcel Kok, and Sonja Kruitwagen. 2016. *Sustainable Development Goals in the Netherlands: Building Blocks for Environmental Policy for 2030.* The Hague: Netherlands Environmental Assessment Agency.

POST2015, Project on Sustainability Transformation beyond 2015. 2016. Prescriptions for Effective Implementation of the Sustainable Development Goals in Japan. Tokyo.

Sachs, Jeffrey D. 2015. *The Age of Sustainable Development.* New York: Columbia University Press.

Stafford-Smith, Mark, David Griggs, Owen Gaffney, Farooq Ullah, Belinda Reyers, Norichika Kanie, Bjorn Stigson, Paul Shrivastava, Melissa Leach, and Deborah O'Connell. 2016. Integration: The Key to Implementing the Sustainable Development Goals. *Sustainability Science.* DOI:10.007/s11625-016-0383-3.

UN DESA, United Nations Department of Economic and Social Affairs. 2015. Global Sustainable Development Report 2015 Edition. United Nations.

UNGA, United Nations General Assembly. 2015. *Transforming Our World: The 2030 Agenda for Sustainable Development.* Draft resolution referred to the United Nations summit for the adoption of the post-2015 development agenda by the General Assembly at its sixty-ninth session. UN Doc. A/70/L.1 of 18 September.

Weitz, Nina, Åsa Persson, Måns Nilsson, and Sandra Tenggren. 2015. Sustainable Development Goals for Sweden: Insights on Setting a National Agenda. Stockholm Environment Institute: Stockholm, Sweden. Working Paper 2015-10.

Annexes

Annex 1: The Millennium Development Goals

Goals and Targets	Indicators for Monitoring Progress
Goal 1: Eradicate extreme poverty and hunger	
Target 1.A: Halve, between 1990 and 2015, the proportion of people whose income is less than one dollar a day	1.1 Proportion of population below $1 (PPP) per day[1] 1.2 Poverty gap ratio 1.3 Share of poorest quintile in national consumption
Target 1.B: Achieve full and productive employment and decent work for all, including women and young people	1.4 Growth rate of GDP per person employed 1.5 Employment-to-population ratio 1.6 Proportion of employed people living below $1 (PPP) per day 1.7 Proportion of own-account and contributing family workers in total employment
Target 1.C: Halve, between 1990 and 2015, the proportion of people who suffer from hunger	1.8 Prevalence of underweight children under 5 years of age 1.9 Proportion of population below minimum level of dietary energy consumption
Goal 2: Achieve universal primary education	
Target 2.A: Ensure that, by 2015, children everywhere, boys and girls alike, will be able to complete a full course of primary schooling	2.1 Net enrolment ratio in primary education 2.2 Proportion of pupils starting grade 1 who reach last grade of primary 2.3 Literacy rate of 15–24 year-olds, women and men
Goal 3: Promote gender equality and empower women	
Target 3.A: Eliminate gender disparity in primary and secondary education, preferably by 2005, and in all levels of education no later than 2015	3.1 Ratios of girls to boys in primary, secondary, and tertiary education 3.2 Share of women in wage employment in the nonagricultural sector 3.3 Proportion of seats held by women in national parliament

Annex 1 (continued)

Goals and Targets	Indicators for Monitoring Progress
Goal 4: Reduce child mortality	
Target 4.A: Reduce by two-thirds, between 1990 and 2015, the under-5 mortality rate	4.1 Under-5 mortality rate 4.2 Infant mortality rate 4.3 Proportion of 1-year-old children immunized against measles
Goal 5: Improve maternal health	
Target 5.A: Reduce by three quarters, between 1990 and 2015, the maternal mortality ratio	5.1 Maternal mortality ratio 5.2 Proportion of births attended by skilled health personnel
Target 5.B: Achieve, by 2015, universal access to reproductive health	5.3 Contraceptive prevalence rate 5.4 Adolescent birth rate 5.5 Antenatal care coverage (at least one visit and at least four visits) 5.6 Unmet need for family planning
Goal 6: Combat HIV/AIDS, malaria, and other diseases	
Target 6.A: Have halted by 2015 and begun to reverse the spread of HIV/AIDS	HIV prevalence among population aged 15–24 years Condom use at last high-risk sex Proportion of population aged 15–24 years with comprehensive correct knowledge of HIV/AIDS Ratio of school attendance of orphans to school attendance of non-orphans aged 10–14 years
Target 6.B: Achieve, by 2010, universal access to treatment for HIV/AIDS for all those who need it	Proportion of population with advanced HIV infection with access to antiretroviral drugs
Target 6.C: Have halted by 2015 and begun to reverse the incidence of malaria and other major diseases	Incidence and death rates associated with malaria Proportion of children under 5 sleeping under insecticide-treated bed nets Proportion of children under 5 with fever who are treated with appropriate anti-malarial drugs Incidence, prevalence, and death rates associated with tuberculosis Proportion of tuberculosis cases detected and cured under directly observed treatment short course

Annex 1 (continued)

Goals and Targets	Indicators for Monitoring Progress
Goal 7: Ensure environmental sustainability	
Target 7.A: Integrate the principles of sustainable development into country policies and programs and reverse the loss of environmental resources Target 7.B: Reduce biodiversity loss, achieving, by 2010, a significant reduction in the rate of loss	Proportion of land area covered by forest CO2 emissions, total, per capita and per $1 GDP (PPP) Consumption of ozone-depleting substances Proportion of fish stocks within safe biological limits Proportion of total water resources used Proportion of terrestrial and marine areas protected Proportion of species threatened with extinction
Target 7.C: Halve, by 2015, the proportion of people without sustainable access to safe drinking water and basic sanitation	Proportion of population using an improved drinking water source Proportion of population using an improved sanitation facility
Target 7.D: By 2020, to have achieved a significant improvement in the lives of at least 100 million slum dwellers	Proportion of urban population living in slums[2]
Goal 8: Develop a global partnership for development	
Target 8.A: Develop further an open, rule-based, predictable, nondiscriminatory trading and financial system Includes a commitment to good governance, development, and poverty reduction—both nationally and internationally Target 8.B: Address the special needs of the least-developed countries Includes: tariff and quota free access for the least-developed countries' exports; enhanced programme of debt relief for heavily indebted poor countries and cancellation of official bilateral debt; and more generous official development assistance for countries committed to poverty reduction	*Some of the indicators listed below are monitored separately for the least-developed countries, Africa, landlocked developing countries, and small island developing states.* Official development assistance Net official development assistance, total and to the least-developed countries, as percentage of OECD Development Assistance Committee donors' gross national income Proportion of total bilateral, sector-allocable official development assistance of OECD Development Assistance Committee donors to basic social services (basic education, primary health care, nutrition, safe water, and sanitation)

Annex 1 (continued)

Goals and Targets	Indicators for Monitoring Progress
Target 8.C: Address the special needs of landlocked developing countries and small island developing states (through the Programme of Action for the Sustainable Development of Small Island Developing States and the outcome of the twenty-second special session of the General Assembly) Target 8.D: Deal comprehensively with the debt problems of developing countries through national and international measures in order to make debt sustainable in the long term	Proportion of bilateral official development assistance of OECD Development Assistance Committee donors that is untied Official development assistance received in landlocked developing countries as a proportion of their gross national incomes Official development assistance received in small island developing States as a proportion of their gross national incomes Market access Proportion of total developed-country imports (by value and excluding arms) from developing countries and least-developed countries, admitted free of duty Average tariffs imposed by developed countries on agricultural products and textiles and clothing from developing countries Agricultural support estimate for OECD countries as a percentage of their gross domestic product Proportion of official development assistance provided to help build trade capacity Debt sustainability Total number of countries that have reached their Heavily Indebted Poor Countries decision points and number that have reached their Heavily Indebted Poor Countries completion points (cumulative) Debt relief committed under Heavily Indebted Poor Countries and MDRI Initiatives Debt service as a percentage of exports of goods and services
Target 8.E: In cooperation with pharmaceutical companies, provide access to affordable essential drugs in developing countries	Proportion of population with access to affordable essential drugs on a sustainable basis

Annex 1 (continued)

Goals and Targets	Indicators for Monitoring Progress
Target 8.F: In cooperation with the private sector, make available the benefits of new technologies, especially information and communications	Fixed-telephone subscriptions per 100 inhabitants Mobile-cellular subscriptions per 100 inhabitants Internet users per 100 inhabitants

1. For monitoring country poverty trends, indicators based on national poverty lines should be used, where available.
2. The actual proportion of people living in slums is measured by a proxy, represented by the urban population living in households with at least one of these four characteristics: (a) lack of access to improved water supply; (b) lack of access to improved sanitation; (c) overcrowding (three or more persons per room); and (d) dwellings made of nondurable material.
Source: Millennium Development Goals Indicators website (visited on September 12, 2014) (http://mdgs.un.org/unsd/mdg/Host.aspx?Content=Indicators/OfficialList.htm)

Annex 2: References to Sustainable Development Goals in the Outcome Document of the 2012 UN Conference on Sustainable Development

Excerpts:

B. Sustainable development goals

245. We underscore that the MDGs are a useful tool in focusing achievement of specific development gains as part of a broad development vision and framework for the development activities of the United Nations, for national priority setting and for mobilisation of stakeholders and resources toward common goals. We therefore remain firmly committed to their full and timely achievement.
246. We recognize that the development of goals could also be useful for pursuing focused and coherent action on sustainable development. We further recognize the importance and utility of a set of SDGs, which are based on Agenda 21 and Johannesburg Plan of Implementation, fully respect all Rio Principles, taking into account different national circumstances, capacities and priorities, are consistent with international law, build upon commitments already made, and contribute to the full implementation of the outcomes of all major Summits in the economic, social and environmental fields, including this outcome document. These goals should address and incorporate in a balanced way all three dimensions of sustainable development and their

inter-linkages. They should be coherent with and integrated in the United Nations development agenda beyond 2015, thus contributing to the achievement of sustainable development and serving as a driver for implementation and mainstreaming of sustainable development in the United Nations system as a whole. The development of these goals should not divert focus or effort from the achievement of the Millennium Development Goals.

247. We also underscore that SDGs should be action-oriented, concise and easy to communicate, limited in number, aspirational, global in nature and universally applicable to all countries while taking into account different national realities, capacities and levels of development and respecting national policies and priorities. We also recognize that the goals should address and be focused on priority areas for the achievement of sustainable development, being guided by this outcome document. Governments should drive implementation with the active involvement of all relevant stakeholders, as appropriate.

248. We resolve to establish an inclusive and transparent intergovernmental process on SDGs that is open to all stakeholders with a view to developing global sustainable development goals to be agreed by the United Nations General Assembly. An open working group shall be constituted no later than the opening of the 67th session of the UNGA and shall comprise of thirty representatives, nominated by Member States through the five UN regional groups with the aim of achieving fair, equitable and balanced geographic representation. At the outset, this open working group will decide on its method of work, including developing modalities, to ensure the full involvement of relevant stakeholders and expertise from civil society, the scientific community and the UN system in its work in order to provide a diversity of perspectives and experience. It will submit a report to the 68th session of the UNGA containing a proposal for sustainable development goals for consideration and appropriate action.

249. The process needs to be coordinated and coherent with the processes considering the post-2015 development agenda. The initial input to the work of the working group will be provided by the United Nations Secretary-General in consultations with national governments. In order to provide technical support to this process and to the work of the working group, we request the UN Secretary-General to ensure all necessary input and support to this work from the UN system including through establishing an inter-agency technical support team and expert panels as needed, drawing on all relevant expert advice.

Reports on the progress of work will be made regularly to the General Assembly.

250. We recognize that progress toward the achievement of the goals needs to be assessed and accompanied by targets and indicators while taking into account different national circumstances, capacities and levels of development.

251. We recognize that there is a need for global, integrated and scientifically based information on sustainable development. In this regard, we request the relevant bodies of the United Nations system, within their respective mandates, to support regional economic commissions to collect and compile national inputs in order to inform this global effort. We further commit to mobilizing financial resources and capacity building, particularly for developing countries, to achieve this endeavor.

Source: UN Conference on Sustainable Development. 2012. The Future We Want. Outcome document from the UN Conference on Sustainable Development. UN General Assembly Res. 66/288. Available at http://www.un.org/en/sustainablefuture/.

Contributors

Dora Almassy is a PhD candidate in the environmental sciences and policy department at the Central European University in Budapest, Hungary. Her thesis research aims to identify the key national governance aspects of international environmental goal-setting and implementation processes and translate these aspects into a sustainability transition management index. In addition, she participates in various international research projects focusing on environmental governance topics. Prior to starting a PhD, she worked as an expert in the environmental financing topic area of the Regional Environmental Center for Central and Eastern Europe. She has a postgraduate degree in business management from Sciences Po, Paris, France, and a master's degree in economics from the University of Miskolc, Hungary.

Steinar Andresen is a research professor at the Fridtjof Nansen Institute of Norway. He has also been a professor at the department of political science at the University of Oslo as well as an adjunct professor at the Pluricourts Center of Excellence, also at University of Oslo. He is a lead faculty member of the Earth System Governance Project and has also been affiliated with the Brookings Institution in Washington DC, Princeton University, University of Washington, and the International Institute for Advanced Systems Analysis in Austria. He has published extensively, particularly on global environmental governance.

Noura Bakkour is project manager at the Institut du Développement Durable et des Relations Internationales (IDDRI), Paris, France. She contributes to IDDRI's efforts to assess and communicate the current state of knowledge regarding financing and implementation of the post-2015 agenda. Previously, Bakkour served as special assistant to Laurence Tubiana, former director at IDDRI. Prior to that, Bakkour held project management and coordination positions at the Earth Institute, Conservation

International, and the Pew Center on Global Climate Change (currently known as C2ES). Bakkour holds a master's degree in public administration from Columbia University, with a focus on environmental science and policy.

Steven Bernstein is a professor in the department of political science and co-director of the Environmental Governance Lab at the Munk School of Global Affairs at the University of Toronto. His research interests include global governance, global environmental politics, international political economy, and international institutions. He is a lead faculty member of the Earth System Governance Project and has consulted on institutional reform for the United Nations. Current major research projects include "Coherence and Incoherence in Global Sustainable Development Governance" (with Erin Hannah) and "Transformative Policy Pathways Towards Decarbonization" (with Matthew Hoffmann).

Frank Biermann is research professor of Global Sustainability Governance with the Copernicus Institute of Sustainable Development, Utrecht University, The Netherlands. He also chairs the Earth System Governance Project, a global transdisciplinary research network that was launched in 2009 and has joined in 2015 the research alliance "Future Earth." Biermann's current research examines options for reform of the United Nations and multilateral institutions, global adaptation governance, Sustainable Development Goals, the political role of science, global justice, and conceptual innovations such as the notion of the Anthropocene. Biermann has authored, co-authored, or edited 16 books, along with numerous articles in peer-reviewed journals and chapters in academic books, with his most recent book being *Earth System Governance: World Politics in the Anthropocene* (MIT Press 2014). He is frequently invited to governmental commissions and panels and has spoken, among other venues, in the UN General Assembly, the European Parliament, and the European Economic and Social Committee.

Thierry Giordano is an agricultural economist at the International Cooperation Centre for Agronomic Research and Development in Montpellier, France. His main fields of expertise are official development assistance, financing for development, and the greening of economies, with a focus on Africa. From 2007 to 2012, he worked for the French Ministry of Foreign Affairs as a technical assistant seconded to the Development Bank of Southern Africa in Midrand, South Africa.

Aarti Gupta is an associate professor in the Environmental Policy Group of the Department of Social Sciences at Wageningen University, The

Netherlands. Her research interests lie in global environmental and sustainability governance, with a focus on the politics of anticipatory risk governance and the role of science, knowledge, and expertise in environmental governance in the issue areas of biotechnology, biodiversity, forests, and climate. Recently, her work has centered on the contested politics of transparency and accountability in environmental governance, with an edited volume on *Transparency in Global Environmental Governance: Critical Perspectives,* published by MIT Press (with Michael Mason, 2014). She is a lead faculty member of the Earth System Governance Project and an associate editor of the journal *Global Environmental Politics*. She holds a PhD from Yale University.

Joyeeta Gupta is professor of environment and development in the Global South at the Amsterdam Institute for Social Science Research of the University of Amsterdam and UN Educational, Scientific and Cultural Organization–IHE Institute for Water Education in Delft. She has published extensively and sits on the scientific steering committees of several national, European, and international projects, including the Earth System Governance Project. Her most recent book, *History of Global Climate Governance,* was published by Cambridge University Press in 2014 and won the Atmospheric Science Librarians International Choice Award for 2014 in its history category.

Peter M. Haas is a professor of political science at the University of Massachusetts, Amherst. He has published extensively on international relations theory, global governance, and international environmental politics. He has received the 2015 UMASS Amherst Award for Outstanding Research and Creative Activities and the 2014 Distinguished Scholar Award of the International Studies Association Environmental Studies Section. He is a lead faculty member of the Earth System Governance Project and has consulted for the UN Environment Programme, UN Commission on Global Governance, and the governments of the United States, France, and Portugal.

Masahiko Iguchi is an assistant professor in the department of international relations at Kyoto Sangyo University, Japan. Prior to his current position, he worked as research associate at the United Nations University Institute for the Advanced Study of Sustainability in Tokyo. Iguchi holds a bachelor's degree in politics and international relations from the University of Essex, a master's degree in international relations from the London School of Economics and Political Science, and a PhD from the Tokyo Institute of Technology.

Norichika Kanie is a professor at the Graduate School of Media and Governance, Keio University. He is also a Senior Research Fellow at United Nations University Institute for the Advanced Study of Sustainability. Before joining Keio, he worked at the Graduate School of Decision Science and Technology, Tokyo Institute of Technology, and the Department of Policy Studies, The University of Kitakyushu. His research focuses on global environmental governance and sustainability. He was the project leader of a research project on Sustainable Development Goals (S-11, FY2013–15), funded by the Ministry of Environment, Japan (Environment Research and Technology Development Fund), of which this book project was a part. Among others he serves on the scientific steering committee of the Earth System Governance Project (a core project of Future Earth), and as co-chair of the Working Party on Climate, Investment and Development of the Organization for Economic Cooperation and Development. In 2009–2010 he was a Marie Curie Incoming International Fellow of the European Commission and visiting professor at SciencesPo, Paris. He holds a PhD in Media and Governance from the Keio University.

Rakhyun E. Kim is an assistant professor of global environmental governance with the Copernicus Institute of Sustainable Development at Utrecht University, The Netherlands. His research explores the complexity of international environmental law from an earth system perspective. Kim serves as book review editor of *Transnational Environmental Law* and is a research fellow with the Earth System Governance Project, an associate fellow at the Centre for International Sustainable Development Law, and a member of the International Union for Conservation of Nature's World Commission on Environmental Law. Kim holds a PhD from the Australian National University. He is the recipient of the 2013 Oran R. Young Prize.

Marcel Kok is environment and development program leader and senior researcher at PBL Netherlands Environmental Assessment Agency, department of Nature and Rural Areas. His research concentrates on governance strategies and scenario analysis of global environmental problems, most recently on mainstreaming biodiversity, bottom-up approaches to global governance, sustainable supply chains, and vulnerability analysis.

Kanako Morita is a researcher at the Bureau of International Partnership, Forestry and Forest Products Research Institute, and a project assistant professor at the Graduate School of Media and Governance, Keio University, Japan. She received her PhD in value and decision science from the Tokyo Institute of Technology in 2010. Her research interests are in environmental policy and governance, including environmental financing.

Måns Nilsson is research director and deputy director at the Stockholm Environment Institute and a part-time professor at the Royal Institute of Technology in Stockholm. Key areas of interest are in low-carbon energy and transport policies, development studies and the 2030 agenda, innovation and transitions, and institutions and governance. In recent years he helped establish the Global Commission on the Economy and Climate and has been closely involved as an advisor on the 2030 Agenda for Sustainable Development and the Sustainable Development Goals to the United Nations, the Organisation for Economic Co-operation and Development, and the European Commission. Måns has slipped more than 40 papers past unsuspecting editors of academic journals and has edited two books. He received his master's degree in international economics from University of Lund, Sweden, and his PhD degree in policy analysis from Delft University of Technology, Netherlands.

László Pintér is a professor in the department of environmental sciences and policy at the Central European University in Budapest, Hungary and Senior Fellow and Associate at the International Institute for Sustainable Development (IISD) in Winnipeg, Canada. Prior to joining the Central European University in 2010, he had been working full time with IISD since 1994, serving as director of IISD's Measurement and Assessment Program between 2003 and 2010. He works worldwide, and his main research areas include the use of knowledge and management tools in sustainable development governance and strategies.

Michelle Scobie is a lecturer and researcher at the Institute of International Relations and the Sir Arthur Lewis Institute for Social and Economic Studies at The University of the West Indies, St. Augustine, Trinidad and Tobago, where she teaches international law, international economic law, and global environmental governance. She is also co-editor of the *Caribbean Journal of International Relations and Diplomacy* and an attorney at law with practice in Trinidad and Tobago and Venezuela. She holds a doctorate in international relations and a bachelor of laws from the University of the West Indies St. Augustine and Cave Hill. Scobie's research areas include global and regional environmental governance trends and challenges, especially in relation to institutional architectures relating to climate change, tourism, sustainable development, marine governance, private governance, environmental ethics, trade, and the environment, particularly from the perspective of developing countries. She is a member of the Earth System Governance Scientific Steering Committee. She has served as a senior economic policy analyst with the Ministry of Finance of the Government of

Trinidad and Tobago and as the first corporate secretary of the Trinidad and Tobago Heritage and Stabilization Fund. She was the recipient of a 2013 Commonwealth Fellowship Award and was a Commonwealth Fellow with the School of International Development of the University of East Anglia, as well as a research fellow at University College, London.

Noriko Shimizu is a researcher at the Institute for Global Environmental Strategies, Japan. Her research includes climate finance and safeguard policies of financial institutions. Shimizu holds a bachelor's degree in political sciences and economics from the Waseda University, Japan and a master's degree in development studies from the University of Bristol, England, and is currently a PhD student at the Tokyo Institute of Technology, Japan.

Casey Stevens is an adjunct faculty member in the department of political science at Clark University in Worcester, Massachusetts. His research focuses on global environmental governance, with a particular emphasis on biodiversity governance and sustainable development. Recent publications have dealt with topics related to global biodiversity politics, including financing and implementation in the green economy era. He is currently working on a book titled *Resilient Governance: Networks for Protecting Changing Ecosystems across Borders*.

Arild Underdal is professor of political science at the University of Oslo and at the Center for International Climate and Environmental Research Oslo, where he works mainly on an eight-year research program known as Strategic Challenges in International Climate and Energy Policy. Most of his research has focused on international cooperation, with particular reference to environmental governance. Other book-format publications include *Environmental Regime Effectiveness: Confronting Theory with Evidence* (with E.L. Miles et al., 2002), *Regime Consequences: Methodological Challenges and Research Strategies* (co-edited with Oran R. Young, 2004), and *The Domestic Politics of Global Climate Change* (co-edited with G. Bang and S. Andresen, 2015). Underdal has served one term (2002–2005) as rector of the University of Oslo and one term as vice rector (1993–1995). Recent international assignments include two terms as chair of the Board of the Stockholm Resilience Centre and two terms as chair of the Science Advisory Committee of the International Institute for Applied Systems Analysis.

Tancrède Voituriez is research officer at the International Cooperation Centre for Agronomic Research and Development and director of the Governance Program at the Institut du Développement Durable et des Relations Internationales, Paris. Following his doctoral research in economics on the instability of commodity markets, he joined the Institut du Développement Durable et des Relations Internationales in 2005 to research the effects of globalization on sustainable development. He has coordinated research projects for the European Commission, the European Parliament, the China Council for International Cooperation on Environment, the UN Food and Agriculture Organization, and the Bill and Melinda Gates Foundation, among others. Since 2010, his work has focused on the conditions for implementing public policies for sustainable development, with a focus on innovative financing.

Takahiro Yamada is a professor of international politics at the Graduate School of Environmental Studies of Nagoya University, Japan. His research has examined the role of knowledge and norms in the creation and evolution of international regimes in areas such as climate change as well as the multi-stakeholder processes that led to the World Bank's socialization of sustainable development norms. He is the author of *Governing an Emerging Global Society*, and numerous articles appearing in the Japan Association of International Relations' journal *International Relations* and the Japanese Society of International Law's *Journal of International Law and Diplomacy*.

Oran R. Young is a professor emeritus at the Bren School of Environmental Science and Management and a research professor at the Marine Science Institute, both at the University of California, Santa Barbara. A longtime leader in research on international environmental governance, his work addresses theoretical questions relating to governance without government and applied issues relating to marine systems, climate change, and the polar regions. His current research focuses on the theme of "sustainability in the Anthropocene: governing complex systems." He has been active for many years in the global environmental research community, and is a lead faculty member of the Earth System Governance Project.

Index